Secure Web-Site Access with Tickets and Message-Dependent Digests

By David I. Donato

Chapter 1 of
Book 7, Automated Data Processing and Computations
Section B, Web Applications

Techniques and Methods 7–B1

U.S. Department of the Interior
U.S. Geological Survey

U.S. Department of the Interior
DIRK KEMPTHORNE, Secretary

U.S. Geological Survey
Mark D. Myers, Director

U.S. Geological Survey, Reston, Virginia: 2008

For product and ordering information:
World Wide Web: http://www.usgs.gov/
Telephone: 1–888–ASK–USGS

For more information on the USGS—the Federal source for science about the Earth, its natural and living resources, natural hazards, and the environment:
World Wide Web: http://www.usgs.gov/
Telephone: 1–888–ASK–USGS

Suggested citation:
Donato, D.I., 2008, Secure Web-site access with tickets and message-dependent digests: U.S. Geological Survey Techniques and Methods, book 7, chap. B1, 53 p., available online only at http://pubs.usgs.gov/tm/7b1/.

Contents

Abstract 1
Introduction 1
    Caveats 1
    Statement of the Problem 1
    Constraints on the Solution 2
        Constraints Related to Usability 2
        Constraints Related to Security 2
        Constraints Related to Economy 2
Survey of the Range of Techniques for Secure Web-Site Access 2
    Three Types of Access Restriction for Web Sites 2
        Restriction by Domain, or IP Address or Subnet 3
        Restriction Through User Authentication 3
        Restriction Through Encryption 3
    State Management and Login Sessions 4
    Digests and Tickets 4
    Comparison of Access-Restriction and State-Management Techniques 5
        Comparison of Desirable Security Traits 5
        Comparison of Resistance to Attacks 6
MDDT: Secure Web-Site Access with Digests and Single-Use Tickets 8
    Client-Server Interactions with this Technique 8
    Strengths and Weaknesses of MDDT 10
A Conjecture about Message-Dependent Digests 11
Summary 11
References Cited 12
Glossary 14
Appendix 1.   List of Abbreviations 20
Appendix 2.   Working JavaScript and PHP code 21
    PHP pages 22
    PHP include files 34
    JavaScript include files 46

Figures

1. Sequence of user, client, and server actions during login with MDDT 9
2. Sequence of user, client, and server actions during transition from one secured
       page to another with MDDT 10

Tables

1. Comparison of desirable Web-site traits enabled by each of seven techniques for Web-site restriction...5
2. Comparison of seven techniques for Web-site restriction with respect to resistance to various kinds of attacks 7

Secure Web-Site Access with Tickets and Message-Dependent Digests

By David I. Donato

Abstract

Although there are various methods for restricting access to documents stored on a **World Wide Web (WWW) site**[1] (a **Web site**), none of the widely used methods is completely suitable for restricting access to **Web applications** hosted on an otherwise publicly accessible Web site. A new technique, however, provides a mix of features well suited for restricting Web-site or Web-application access to authorized users, including the following: secure user **authentication**, tamper-resistant **sessions**, simple access to user **state** variables by **server**-side applications, and clean session terminations. This technique, called message-dependent digests with tickets, or **MDDT**, maintains secure user sessions by passing single-use **nonces (tickets)** and **message-dependent digests** of user credentials back and forth between **client** and server. Appendix 2 provides a working implementation of MDDT with **PHP** server-side code and JavaScript client-side code.

Introduction

This publication presents a technique for providing secure access to selected documents on a World Wide Web site (a Web site). The technique enables a **webmaster** to secure a subset of the digital files on a Web site so these files can be accessed only by **authorized** users.

The technique presented here uses **message digests** that are message-dependent to enable users to identify (authenticate) themselves to a Web site without passing their user credentials over the network during the authentication process. The technique also uses complex character strings called **authentication tickets** to create **login sessions** for **authenticated** users. Though message digests and ticket-like nonces have been used previously in other approaches to user authentication and **state management** for Web applications, the particular combination of message-dependent message digests and single-use tickets (nonces) presented here has unique features that makes it more suitable than earlier approaches for

---

[1]Throughout this article the first appearance of each term defined in the glossary or appendix 1 is identified in **boldface**.

some applications. For clarity and ease of reference, this publication refers to this new technique as MDDT (**m**essage-**d**ependent **d**igests with **t**ickets). The Appendix provides computer code for a working implementation of MDDT.

Caveats

MDDT is not suitable for all applications and is presented without guarantees. Careful webmasters and Web-application developers may apply MDDT directly or adapt it to their needs. Like any other security technique related to **information technology (IT)**, MDDT should be applied circumspectly, in consultation with suitable IT security specialists, as one of many components of a total IT security solution.

The author is not an expert in information-technology security. He is an occasional Web-application developer sharing his practical experience from developing this technique to enable secure, restricted access to a USGS Web site by non-USGS collaborators. Mounting concerns within the U.S. Department of the Interior about the security risks associated with publicly accessible Web sites and the lack of available infrastructure for **extranets** (Open Web Application Security Project, 2002; Venners, 2006) prompted the development of MDDT. Although the Glossary defines technical terms as they are used to describe MDDT, it is not a substitute for substantial familiarity with **HyperText Transfer Protocol (HTTP)** (Network Working Group, Request for Comments (**RFC**) 2616, 1999b), with other **Internet** protocols, and with the security problems of computer networks and Web sites. The "References Cited" section of this publication lists selected authoritative publications that cover the required background for this publication in the areas of general IT and Web-application security (Garfinkel and others, 2003; Pastore and Dulaney, 2004; Skoudis with Liston, 2006; Stein and Stewart, 2002).

Statement of the Problem

MDDT was designed to solve the following problem: How can access to selected Web applications, files, and forms hosted on a publicly accessible Web site be conveniently and economically restricted to a small group of authorized users, subject to reasonable constraints on cost and ease of use?

This problem, which was encountered in the development of an extranet-like facility to enable collaborative use of computer models, is not the same as the problem encountered by banks and online merchants who need to protect the contents of their online transactions from disclosure to third parties. In the problem addressed here, it is only the ability to run Web applications that must be restricted by limiting access to operational Web forms; the files exchanged during the operation of these applications are assumed to be in no need of protection from disclosure.

Constraints on the Solution

The constraints that were applied in MDDT design in order to provide acceptable usability, security, and economy are listed separately below for each of these categories:

Constraints Related to Usability

- An authorized user should be able to access information from essentially any location (including offices, homes, and hotel rooms) where broadband or dial-up network access is available.

- An authorized user should be able to use any of various browsers running under Windows operating system,[2] Linux, UNIX, MacOS, or any other network-capable operating system.

- An authorized user should not have to download or install special software.

- An authorized user should not have to enter authentication credentials (such as user name and password) more than once per session.

- There should be no perceptible delays caused by the access-restriction solution.

Constraints Related to Security

- The solution should be able to resist determined **attacks**.[3]

- The solution should not rely solely on the validity of **IP addresses** (Network Working Group, RFC 1518, 1993) or **Domain Name System (DNS)** aliases (Network Working Group, RFC 1034, 1987).

- The solution should not transmit user credentials over a network in clear text.

- A restricted document should not be accessible through **Uniform Resource Locator (URL) guessing**.

Constraints Related to Economy

- Users (clients) who have general Internet access should not have to purchase additional software or hardware in order to access the restricted resources.

- The webmaster or Web-application developer should not have to purchase additional hardware or software in order to implement the solution.

- The solution should not entail any periodic licensing fees.

- The solution should be simple enough to be implemented for 10 Web pages in under 2 hours.

Survey of the Range of Techniques for Secure Web-Site Access

To put MDDT into perspective, this section surveys the kinds[4] of techniques available for securing selected files on publicly accessible Web sites by restricting access (Venners, 2006). This section also discusses the closely related topics of state management and login sessions. The explanation of message digests and tickets leads into a comparison of MDDT capabilities with the capabilities of selected alternative techniques. (The detailed processing steps of MDDT are explained later in a separate section.) This section provides a basis for understanding the kinds of applications for which MDDT is suited and those for which it is not.

Three Types of Access Restriction for Web sites

There are three general methods (Stein and Stewart, 2002) for restricting access to selected files on an otherwise publicly accessible Web site: (1) by **domain**, or IP address or **subnet**; (2) through user authentication; or (3) through encryption. These methods may be applied separately or in combination. Each has strengths and weaknesses that affect its suitability for any specific application.

---

[2]Windows is a registered trademark of Microsoft Corporation in the United States and other countries.

[3]What constitutes a determined attack varies from one situation to another. As applied to the design of MDDT, the practical intent of this constraint is that the solution should not have any known weaknesses that might be exploited without at least several hours of effort on the part of an attacker.

---

[4] The author is not aware of the existence of any other techniques equivalent to MDDT. A search of the World Wide Web for software patents related to secure Web-site access found two partially relevant patents (Bui and others, 2002; Sasmazel and Schneider, 2000). MDDT does not, however, infringe on either patent.

Restriction by Domain, or IP Address or Subnet

Widely used Web server software, such as Apache and Microsoft® Internet Information Services (IIS), provides simple ways to restrict access to **Web** files based on the client domain, or the IP address, or the IP subnet. This form of restriction is suitable when all authorized users have persistent, fixed IP addresses or when authorized users only access the Web site from client workstations located within a fixed domain; this form of restriction is not suitable, however, when IP addresses for authorized users are unpredictable, such as when users attend conferences, work from hotels, or use **Internet Service Providers** (**ISP**s) that do not provide users with fixed IP addresses. There is a **workaround** when users' IP addresses are unpredictable or changeable: the user's IP address (or DNS domain) can be captured by software, as the user logs onto the restricted Web application; then it can be placed in an **access control list** (**ACL**) of allowed IP addresses. This workaround is not, however, directly supported by widely used Web server software; it requires the development of server-side software (scripts or programs) to handle user logons and to manage the IP addresses in the ACL.

Restriction Through User Authentication

There are two kinds of user authentication directly supported by some Web servers and some Web browsers: HTTP Basic Authentication and HTTP Digest Authentication (Network Working Group, RFC 2617, 1999c). These two HTTP authentication methods enable restrictions on files at the same levels of granularity allowed for other Web-server access restrictions (generally per directory, per Web site, or per **virtual host**). These two methods differ in that under basic authentication, passwords are passed over the network in clear text, while under digest authentication, only a digest of the user password is sent over the network (World Wide Web Consortium, 2001; Apache HTTP Server Project, 2005). (Please see "Digests and Tickets" below for an explanation of message digests and their use in user authentication.) Basic authentication is widely supported by Web servers and Web browsers; support for digest authentication, however, remains spotty (Apache Week, 1996; Laurie and Laurie, 1999; Apache HTTP Server Project, 2006).

Web-application developers can, and sometimes must, write their own user-authentication routines (IBM, 2006). Custom authentication code is necessary when user identity has complex effects within a Web application, such as when users are allowed to resume interrupted processes or set personal preferences or save individual documents. Custom code for user-authentication routines may also be made necessary by application-specific requirements related to state management. (Please see "State Management and Login Sessions" below.)

Restriction Through Encryption

Encryption reversibly converts a file into an unreadable or unusable form through the use of **cryptographic ciphers**. If encrypted with suitably secure ciphers, files transferred over a network are secure from **eavesdropping** (Barrett and others, 2005). Some Web applications apply encryption only to user credentials in order to protect them from eavesdropping or interception while in transit, while other Web applications encrypt all transmissions. When the volume of information encrypted and transmitted is relatively small, **encryption** of all transmissions is feasible; but when large files must be transmitted, encrypting them for transmission and decrypting them on receipt may be unacceptably time consuming.

There are several established methods for using encryption to restrict access on the Web. **HTTPS** (**HyperText Transmission Protocol Secure**) is widely used by banks, online shopping sites, and others to provide end-to-end (client-to-server) encryption; HTTPS[5] transmits files with HTTP over **Secure Sockets Layer** (**SSL**), or **Transport Layer Security** (**TLS**), or **OpenSSL** (Network Working Group, RFC 2246, 1999a; Network Working Group, RFC 2818, 2000). Alternatively, programmers can apply the same kinds of encryption in custom software by developing code to work with the OpenSSL library (Viega and others, 2002; PHP Group, 2007). A less widely used protocol, **S-HTTP** (**Secure HTTP**), also provides end-to-end encryption, though just for particular messages rather than for all HTTP request-response exchanges (Network Working Group, RFC 2660, 1999d). The computational overhead and the setup costs of HTTPS and S-HTTP can be avoided when only a set of seldom changed files must be sent over the network but protected from disclosure to unauthorized persons; in this case, the sensitive files may be served from a Web site in encrypted form as long as only authorized users possess the keys needed to decrypt them. Yet another use of Web-site restriction through encryption requires that Web sites with sensitive information be placed behind firewalls and then be made accessible to remote users only through a **Virtual Private Network** (**VPN**) that encrypts all transmissions from one network (such as a network operated by a user's ISP) to the Web site's firewalled home network. A widely used encryption protocol for VPNs is **IPsec** (Network Working Group, RFC 2401, 1998).

In addition to its use in protecting information in transit, encryption is also used to protect stored information. When authentication data (especially passwords) are stored in encrypted files, these files must be protected from unauthorized access to prevent attackers[6] from running them through

---

[5]Under HTTPS (and other **client-server** protocols involving encryption), both client and server can achieve reasonable assurance of the identity of the other party through third-party certificates provided through the **Public Key Infrastructure (PKI)** (Network Working Group, RFC 3647, 2003; Viega and others, Chapter 3, 2003).

[6]For simplicity, any attempt to interfere with an information-systems resource, or any attempt at unauthorized access, is referred to as an attack, and the agent of the attack is called an attacker (Skoudis with Liston, 2006).

password-**cracking** routines. Since some attackers are willing to run password-cracking software for weeks in order to decrypt a password file, best security practices require that passwords be changed frequently under any circumstances, and immediately following any confirmed or suspected incident of unauthorized access to a password file (Garfinkel and others, 2003; Skoudis with Liston, 2006).

State Management and Login Sessions

HyperText Transfer Protocol (HTTP), the set of rules that governs file transport between **Web clients** and **Web servers**, is a **stateless** protocol. An HTTP exchange between a client (a Web browser) and a server generally consists of little more than a request and a response: first, a file is requested by the client; then the file or a reply message is sent; and, finally, both browser and Web server close out the exchange.[7] Because HTTP is stateless, the notion of a user session for a Web site is not defined at the level of the HTTP transfer protocol. User sessions (if there are to be any) must be defined by the Web application.

The concept of the user state or client state is broader than, and subsumes, the idea of a user login session. State can be managed without using login sessions, but not the other way around. A Web application that defines and manages its own state and the state of each active user does not necessarily have to use login sessions; an application must, however, create and manage state variables in order to make user login sessions possible—and doing so is a form of state management.

There are a number of frameworks available for the development and operation of Web applications that provide high-level features to simplify the job of state management for Web-application developers. Among the most widely used of these frameworks are Active Server Pages (ASP); ASP.NET; Java Server Pages (JSP); PHP Hypertext Preprocessor (PHP); and Z Object Publishing Environment (Zope). The state-management facilities provided by these frameworks can be used in conjunction with any of the three types of access restriction in order to define access sessions; when access is, in particular, restricted through user authentication, these state-management facilities can be used to define not just access sessions, but login sessions as well. Although development frameworks are beneficial for some applications, for others they are too confining; in the latter cases, Web-application developers must create custom techniques for managing state and defining login sessions (as was done in developing MDDT).

Digests and Tickets

A message digest is a compact string or a number reproducibly derived from any ordered collection of bytes, such as a character string, a digital document, or any other computer file. It is called a digest[8] because it is usually much smaller than its source. Even when the source collection includes millions or billions of bytes, a message digest typically consists of 30 or 40 bytes at most. Ideally, finding the original collection of bytes from the digest, with both digest and the digest **algorithm** known, should be difficult in the extreme, and the probability that any two messages chosen at random have the same digest should be small (certainly less than one in a million, and preferably far less). Assuming there is no true reverse algorithm and the forward algorithm is known, the only way to identify a message from its digest is to repeatedly guess at the original message and compute a digest for each guess until a message is found that produces the same digest. (Even then, because digests do not have to be unique, a message found through this sort of guesswork might not be the original message.) For a message consisting of 20 characters drawn from the 72 characters made up of the upper- and lower-case letters, the digits, and 10 special characters, the number of possible messages is 72^{20}, or more than 1.4×10^{37}. At a trillion guesses per second, it would still take more than 4×10^{17} years to work through all possibilities; therefore, unless a message is short and easy to guess, identification of the source message from a digest created with a well-constructed digest algorithm is computationally infeasible.

Digest authentication would not be any more secure than just sending the user name and password over the network in clear text if the digest were based on user credentials alone, because in either case, an eavesdropper could capture the digest (or the credentials) and use it (or the credentials) to access the restricted application. To make message digests effective in securing access to a Web site, it is necessary to add a nonce to the user credentials so that a different digest will be generated for each login session. The Web application generates the nonce and provides it to the client, which then computes a message digest of user credentials concatenated with the nonce and returns this digest to the Web application for user authentication. Since the Web application has access to the same data as the client (including the nonce it just sent and a collection of all valid user credentials), the application can generate a matching message digest to verify the identity of the user. In MDDT, the server generates a new nonce for every client request; this makes each digest essentially unique and user login sessions secure. Only a user who knows the credentials with which a session was initiated and who is in receipt of the latest nonce can successfully request another restricted file from the Web server.

---

[7]This is a simplification. Web servers often wait several minutes before dismantling the lower-level socket connection for each client because there may be prompt follow-on requests. Keeping a socket connection alive does not, however, preserve any information about the client (user state), which could affect Web-server responses to future HTTP requests from the same client. These socket-level keepalives are features of low-level network connection management, not of state management.

[8]A message digest is sometimes alternatively called a **hash**. Federal Information Processing Standards mandate the use of the SHA-1 or related hash (digest) algorithms for any Federal application requiring a secure hash (digest) algorithm (National Institute of Standards and Technology, 2002).

Table 1.   Comparison of desirable Web-site traits enabled by each of seven techniques for Web-site restriction.

[In this table, a plus sign (+) in a cell with a light-green background indicates that the technique identified in the column heading provides a significant measure of the trait identified by the row label on the left; an em dash (—) indicates that the technique does not provide the trait in significant measure. HTTP Digest Authentication, for example, does enable restriction of access to authenticated users and, therefore, maintains a significant level of document confidentiality; it does not, however, significantly ensure document integrity, nonrepudiation of access, nor session continuity. The ratings in this table provide, at best, a rough overall comparison among techniques]

Trait	Technique used for Web-site security and state management						
	Message-dependent digests with tickets	HTTP basic authentication	HTTP digest authentication	Restriction by IP address	IPsec	HTTPS	PHP state management
Authentication	+	+	+	—	+	+	—
Confidentiality	+	+	+	+	+	+	—
Integrity	+	—	—	—	+	+	—
Nonrepudiation	+	—	—	—	—	+	—
Session continuity	+	—	—	—	—	—	+

A more natural term than nonce is **ticket**[9] (or authentication ticket) because **ticket** is a commonly used word that connotes something arbitrary, ad hoc, lacking in intrinsic value, and time-limited. The MDDT nonce is called the "ticket" throughout the remainder of this publication.

Comparison of Access-Restriction and State-Management Techniques

A two-way comparison of MDDT with six other techniques for restricting access to Web sites serves both to clarify the practical strengths and weaknesses of MDDT, and to establish a basis for choosing among MDDT and its alternatives. The first comparison rates the seven techniques on the basis of their desirable traits (the security features each technique provides). The second comparison evaluates the relative resistance of each of the seven techniques to each of 13 widely recognized kinds of attacks[10] (Open Web Application Security Project, 2007).

Comparison of Desirable Security Traits

The desirable traits of access-restriction techniques evaluated for comparison are:

1. **Authentication** – Ability to restrict access to authorized users only.

2. **Confidentiality** – Ability to prevent access to Web-site files by unauthorized users.

3. **Integrity** – Ability to assure users that there has been no tampering with the documents they retrieve from a Web site.

4. **Nonrepudiation** – Ability to conclusively associate a specific user with specific Web-site activities and actions.

5. **Session Continuity** – Ability to maintain a user's context and application state throughout a login session.

Table 1 summarizes a comparison of the occurrence of these five desirable traits among seven techniques for restricted Web-site access. In comparison to the other six techniques, MDDT stands out as a more complete security solution than any other technique by itself, although there are potential combinations of other techniques that would be equally complete. Because MDDT does not transmit user credentials over the network and establishes user sessions that clearly associate individual users with their activities during tamper-resistant login sessions, it exhibits to a significant degree all five desirable traits: authentication, confidentiality, integrity, nonrepudiation,[11] and session continuity. Although HTTPS provides neither user-authentication nor state-management services at the application level, other state-management and user-authentication facilities can be used in conjunction with HTTPS, and when even simple login procedures are used under HTTPS, the login credentials are

---

[9]The term **ticket** is used with similar meaning in the Kerberos network authentication protocol.

[10]Cross-site scripting attacks have intentionally not been included in the list of attacks. Although cross-site scripting attacks are a potentially serious problem for many Web applications, neither MDDT nor any of the other techniques compared are susceptible to this type of attack. Unlike **denial-of-service** attacks, which are included because they could become a problem if some of the access-restriction techniques are misused, cross-site scripting attacks simply will not affect these access-restriction techniques under any circumstances.

[11]MDDT's control is not strong enough to establish nonrepudiation in any legally enforceable sense. MDDT's combination of digest authentication integrated with a session-control mechanism does, however, provide notably stronger recordation and proof of user actions than the other techniques.

protected in transit by HTTPS encryption. Therefore, HTTPS[12] exhibits all traits, except for session continuity, in the broad sense that it is compatible with simple implementations of those features it does not provide directly. IPsec is effective in securing communications between one network and another, but it does not otherwise restrict access to any Web site or Web application by Intranet users; therefore, IPsec exhibits just the traits of authentication, confidentiality, and integrity related to Web-site access, but only if all persons with access to the private network are also authorized users of the restricted Web site or Web application hosted on this private network. HTTP Authentication (both Basic and Digest) supports authenticated access and provides confidentiality, but neither form of HTTP Authentication creates clear-cut sessions that convincingly associate a user with Web-site activities. Of the techniques compared, IP-address restriction provides the weakest limit on Web-site access, restricting access only to any user of any computer workstation with an IP address in the allowed range. Even though PHP state management addresses only the problem of state management and user sessions, not restriction of access (Lerdorf and Tatroe, 2002), PHP state management[13] is included in the comparison because of its potential use in composite security solutions as a complement to other techniques.

Some of the techniques (notably HTTPS and IPsec) are more robust and better tested than MDDT, and are clearly superior for many applications. The comparison here is only concerned with the features provided and with the problems addressed by each technique, not with the stability, resilience, or extent of testing of the various techniques.

Comparison of Resistance to Attacks

The 13 widely recognized kinds of attacks[14] chosen for comparison are as follows:

1. **Account harvesting** – Gathering user names or passwords or both for later break-in attempts.

2. **Credentials guessing** – Repeated, usually automated, login attempts with guessed user names and passwords.

3. **Data spoofing (tampering)** – Unauthorized modification of computer files or transmissions.

4. **Denial of service** – Overloading a resource in order to make it unavailable.

5. **Eavesdropping** – Secretly obtaining information transmitted between two other parties without stopping or detectably slowing transmission.

6. **Interception** – Secretly obtaining information intended for another recipient and preventing receipt by the intended recipient.

7. **IP spoofing** – Making transmissions on the network as if they came from an IP address other than the true IP address of the source.

8. **Person-in-the-middle** – Secretly modifying transmissions between two other parties as they take place. (This attack is also known as **man-in-the-middle**, or **MITM**.)

9. **Replay** – Use of captured credentials to obtain access at a later time.

10. **Session cloning (client-side masquerading)** – Secretly or deceptively taking over as a session client using intercepted session variables.

11. **Session hijacking (server-side masquerading)** – Secretly or deceptively redirecting a session client to a different server.

12. **Social engineering** – Techniques of deception used in social interactions to obtain information and credentials for accessing restricted Web sites.

13. **URL guessing** – Using known URLs to guess a URL for a Web page in order to gain unauthorized access.

Table 2 summarizes the ability of the seven access-restriction techniques to resist each of 13 kinds of attacks.

IP spoofing can overcome access restrictions based solely on IP addresses. IP spoofing alone is not, however, effective against any of the other techniques, although IP spoofing may be used in conjunction with other attacks to hide the identity of the attacker.

Although Web-server availability can be compromised by the overloading of a Web application, it is more usual for denial-of-service (**DoS**) attacks to take place below the application level. Whether a Web site or application is a potential target for an application-level DoS attack depends on the nature of the site or application; therefore, the resistance of the various Web-site security and state-management techniques to DoS attacks has not been specifically evaluated. DoS attacks, however, have been included in the comparison to acknowledge the potential for Web-application-level DoS attacks in the event that vulnerabilities are discovered in the access-restriction techniques.

Social engineering can be effective against almost any technique, except for those that are beyond user (or attacker) control; therefore, six of the seven techniques are not resistant to social engineering, but restriction by IP address does offer some resistance. Admittedly, IP spoofing, in conjunction

---

[12]In general, HTTPS does not work with virtual hosts, so it may not be suitable for Web applications that share physical Web-server hosts.

[13]PHP State Management serves in these comparisons as a representative of the state-management facilities provided by various Web-development frameworks.

[14]The list of 13 kinds of attacks is not an exhaustive list of all possible attacks against Web sites. The list is intended only to include enough kinds of attacks to allow effective comparison among the seven Web-site restriction techniques for the practical purpose of selecting specific techniques for specific applications.

Table 2. Comparison of seven techniques for Web-site restriction with respect to resistance to various kinds of attacks.

[In this table, a plus sign (+) in a cell with a light-green background indicates that the technique identified in the column heading is largely resistant to the kind of attack identified by the row label on the left; an em dash (——) indicates that the technique is nonresistant or only weakly resistant. Cells are marked N/A (not applicable) when resistance to a particular kind of attack is generally not affected by the technique but could be affected if the technique is misapplied. The ratings in this table provide, at best, a rough overall comparison among techniques]

Technique used for Web-site security and state management

Kind of attack	Message-dependent digests with tickets	HTTP basic authentication	HTTP digest authentication	Restriction by IP address	IPsec	HTTPS	PHP state management
Account harvesting	+	—	+	N/A	—	+	—
Credentials guessing	+	—	+	N/A	—	—	—
Data spoofing (tampering)	—	—	—	—	—	+	—
Denial of service	N/A	N/A	N/A	N/A	N/A	N/A	N/A
Eavesdropping	+	—	—	—	—	+	—
Interception	—	—	—	—	—	+	+
IP spoofing	N/A	N/A	N/A	—	N/A	N/A	N/A
Person-in-the-middle	—	—	+	—	—	—	—
Replay	+	—	+	—	—	—	+
Session cloning (client-side masquerading)	+	N/A	+	—	—	—	—
Session hijacking (server-side masquerading)	+	—	+	—	—	—	—
Social engineering	—	—	—	+	—	—	—
URL guessing	+	+	+	—	—	—	—

with a social-engineering effort to determine valid IP address ranges, can defeat IP-address restrictions. Because MDDT makes use of the client IP address to track user login sessions, IP address restrictions can be easily added to MDDT.

Account harvesting is difficult for attackers when little or no information about user accounts can be seen either by eavesdropping on logins with a **packet sniffer** or by attempting multiple logins to observe changes in the server response. For this reason MDDT, HTTP Digest Authentication, and HTTPS are resistant to account harvesting. Because intercepted digests are of essentially no use in guessing or **cracking** user credentials (user names, passwords, and pass phrases), and because an attacker must guess all user credentials exactly in order to gain access, MDDT and HTTP Digest Authentication also provide good protection against credentials guessing. The resistance of HTTPS to credentials guessing is moot because HTTPS does not itself provide application-level user-authentication features; resistance to credentials guessing when HTTPS is in use depends on the resistance of the associated application-level user-authentication technique.

The best protection against tampering, eavesdropping, or **interception** of transmissions is the strong end-to-end encryption provided by HTTPS. None of the other solutions provides significant protection against these attacks, except that the techniques that provide strong session continuity (MDDT and PHP State Management) do protect against interception because breaks in sessions are obvious to users and are likely to trigger investigations.

Six of the seven of the techniques are vulnerable to person-in-the-middle (**PITM**) attacks, which are usually carried out by interposing a **proxy** either on the client's network or on the server's network. Whether on the client side or the server side, a PITM attack entails redirection of transmissions to the proxy. HTTP Digest Authentication is resistant to PITM attacks because neither client browser nor Web server provides a mechanism for revealing the user credentials used to create digests. MDDT, despite its use of digests to protect user credentials, is theoretically vulnerable to PITM attacks because an attacker could add a client-side script in order to cause user credentials to be sent in clear text as parameters when the user clicks on a link.[15]

Digest authentication and strong state management protect against replay attacks by generating digests and state variables that can only be used once, and then only if used promptly. None of the other techniques resist replay attacks, though replay is unlikely to be an application-level threat when the application operates over IPsec or HTTPS. Session

cloning and session hijacking are unlikely when digest authentication is in use because an attacker lacks the user credentials needed to clone a session (masquerade as the client) and would not be able to determine user credentials by hijacking a session (masquerading as the server). Other than MDDT and HTTP Digest Authentication, none of the techniques protects against session cloning or hijacking. MDDT provides additional protection against session cloning by tracking user sessions with the IP address; an attempt to take over an MDDT session with a new client IP address would immediately be rejected at the Web server.[16]

Only the three techniques that specifically provide user authentication (MDDT and HTTP Basic and Digest Authentication) prevent access to Web pages by URL guessing. With these techniques, an unauthorized user will be denied access to a guessed page (but may determine that a page for the guessed URL does exist).

MDDT: Secure Web-Site Access with Digests and Single-Use Tickets

MDDT uses a single integrated mechanism to achieve combined user authentication and login-session continuity. Working computer code provided in appendix 2 implements MDDT using JavaScript on the client side and PHP on the server side, but the underlying technique of MDDT can be implemented with any of a number of other client-side and server-side programming languages.

Client-Server Interactions with this Technique

Figure 1 describes the process by which a client logs in to begin a session. Figure 2 describes the authentication mechanism underlying continuing access to restricted documents and other information resources during a login session. The actual code used to generate tickets and digests is the best source for understanding in detail how digests and tickets are created. Please keep the following points in mind when reviewing code and examining figures 1 and 2:

- Tickets depend on the current value of the server's clock; this makes it nearly impossible for an attacker to guess what ticket the server will generate next.

- A new ticket is generated for each Web page requested, making tickets perishable.

- Message digests are generated using a procedure (algorithm) that modifies its own operation based on the message to be digested.

---

[15]HTTP digest authentication and MDDT differ in that HTTP digest authentication stores client-side user credentials in variables located within the browser process itself, but MDDT stores credentials within browser windows and client-side documents. Variables within the Web-browser process are generally not accessible to incoming client-side scripts; however, browser-window and document variables are accessible. Browser-process variables are much more secure than window or document variables. Cookies, whether stored on disk or in memory, are no more secure than window and document variables because cookies are also accessible to client-side scripts.

[16]Because MDDT uses client IP addresses for session tracking, multiple users cannot simultaneously connect to the same MDDT application with identical IP addresses. This would be a limitation when Network Address Translation (NAT) or proxy servers are in use because these allow multiple users to share a single IP address.

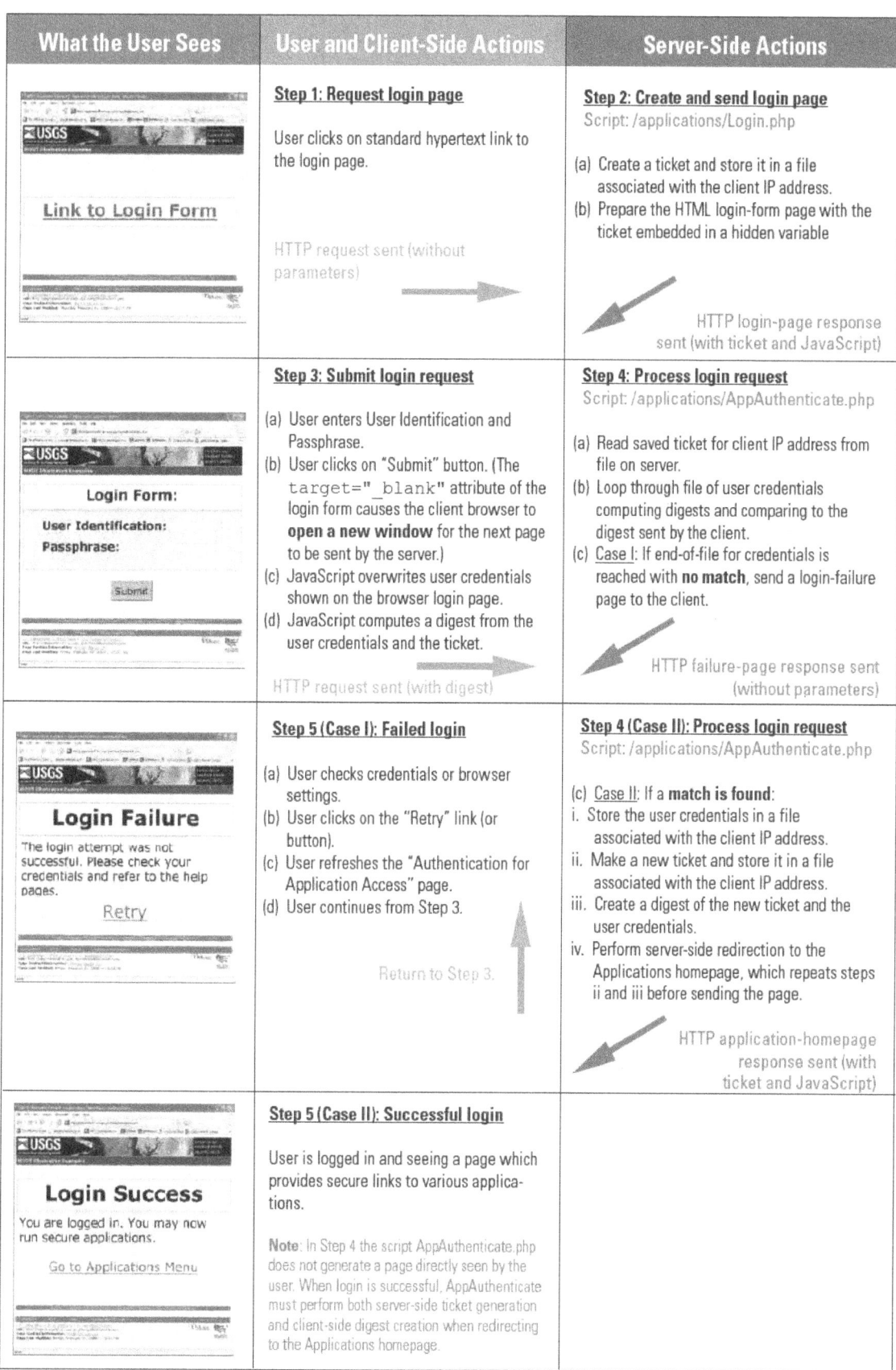

What the User Sees	User and Client-Side Actions	Server-Side Actions
Link to Login Form	**Step 1: Request login page** User clicks on standard hypertext link to the login page. HTTP request sent (without parameters)	**Step 2: Create and send login page** Script: /applications/Login.php (a) Create a ticket and store it in a file associated with the client IP address. (b) Prepare the HTML login-form page with the ticket embedded in a hidden variable HTTP login-page response sent (with ticket and JavaScript)
Login Form: **User Identification:** **Passphrase:** Submit	**Step 3: Submit login request** (a) User enters User Identification and Passphrase. (b) User clicks on "Submit" button. (The `target="_blank"` attribute of the login form causes the client browser to **open a new window** for the next page to be sent by the server.) (c) JavaScript overwrites user credentials shown on the browser login page. (d) JavaScript computes a digest from the user credentials and the ticket. HTTP request sent (with digest)	**Step 4: Process login request** Script: /applications/AppAuthenticate.php (a) Read saved ticket for client IP address from file on server. (b) Loop through file of user credentials computing digests and comparing to the digest sent by the client. (c) Case I: If end-of-file for credentials is reached with **no match**, send a login-failure page to the client. HTTP failure-page response sent (without parameters)
Login Failure The login attempt was not successful. Please check your credentials and refer to the help pages. Retry	**Step 5 (Case I): Failed login** (a) User checks credentials or browser settings. (b) User clicks on the "Retry" link (or button). (c) User refreshes the "Authentication for Application Access" page. (d) User continues from Step 3. Return to Step 3.	**Step 4 (Case II): Process login request** Script: /applications/AppAuthenticate.php (c) Case II: If a **match is found**: i. Store the user credentials in a file associated with the client IP address. ii. Make a new ticket and store it in a file associated with the client IP address. iii. Create a digest of the new ticket and the user credentials. iv. Perform server-side redirection to the Applications homepage, which repeats steps ii and iii before sending the page. HTTP application-homepage response sent (with ticket and JavaScript)
Login Success You are logged in. You may now run secure applications. Go to Applications Menu	**Step 5 (Case II): Successful login** User is logged in and seeing a page which provides secure links to various applications. **Note**: In Step 4 the script AppAuthenticate.php does not generate a page directly seen by the user. When login is successful, AppAuthenticate must perform both server-side ticket generation and client-side digest creation when redirecting to the Applications homepage.	

Figure 1. Sequence of user, client, and server actions during login with MDDT.

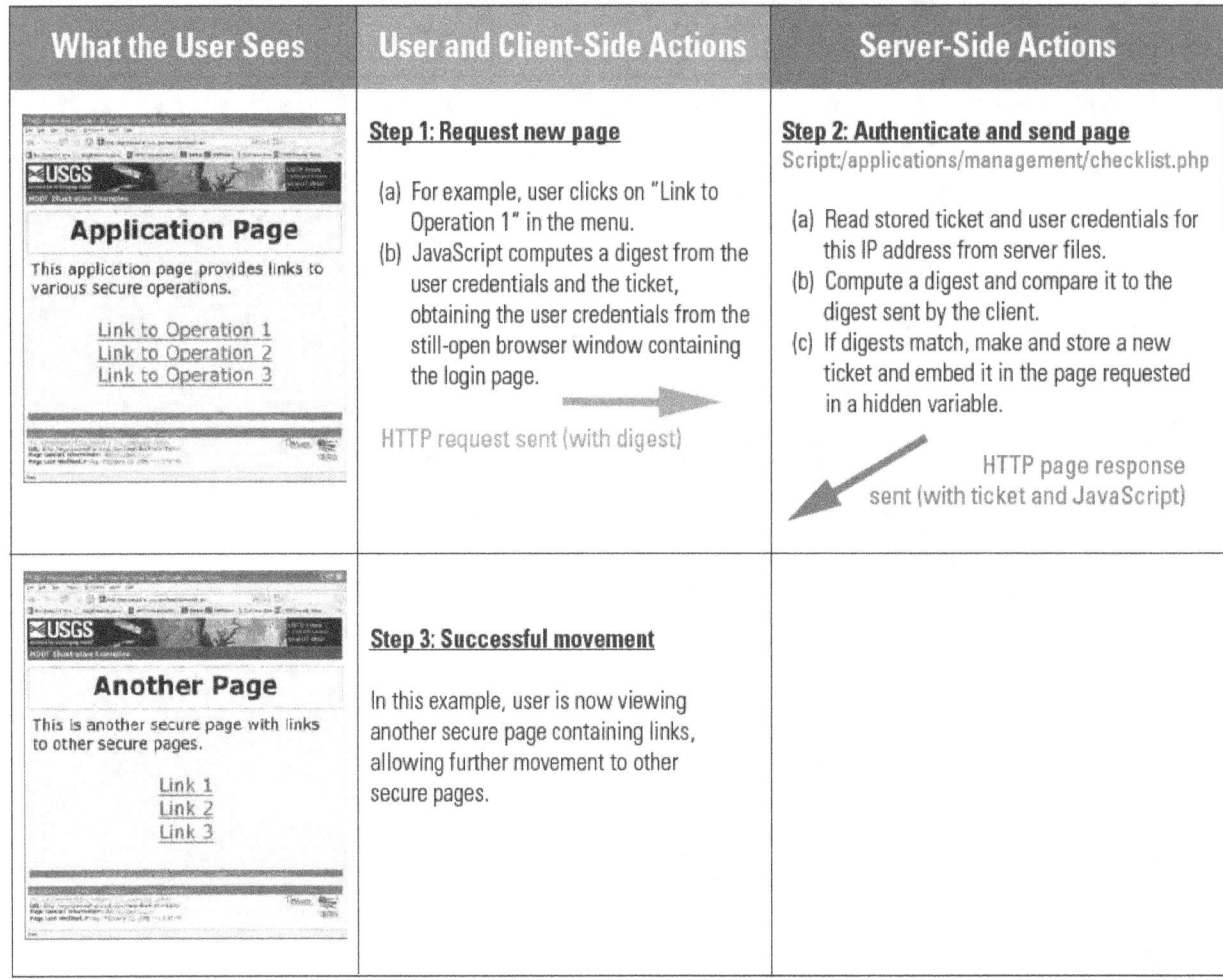

What the User Sees	User and Client-Side Actions	Server-Side Actions
Application Page This application page provides links to various secure operations. Link to Operation 1 Link to Operation 2 Link to Operation 3	**Step 1: Request new page** (a) For example, user clicks on "Link to Operation 1" in the menu. (b) JavaScript computes a digest from the user credentials and the ticket, obtaining the user credentials from the still-open browser window containing the login page. HTTP request sent (with digest)	**Step 2: Authenticate and send page** Script:/applications/management/checklist.php (a) Read stored ticket and user credentials for this IP address from server files. (b) Compute a digest and compare it to the digest sent by the client. (c) If digests match, make and store a new ticket and embed it in the page requested in a hidden variable. HTTP page response sent (with ticket and JavaScript)
Another Page This is another secure page with links to other secure pages. Link 1 Link 2 Link 3	**Step 3: Successful movement** In this example, user is now viewing another secure page containing links, allowing further movement to other secure pages.	

Figure 2.   Sequence of user, client, and server actions during transition from one secured page to another with MDDT.

- The message-dependent character of the digest algorithm, by conjecture, prevents or severely hinders the development of a computationally tractable reverse algorithm.

Strengths and Weaknesses of MDDT

MDDT has the following strengths:

- The digest algorithm for an implementation can be easily changed through a minor revision of the software with low risk of introducing bugs into other parts of the MDDT code.

- When high-strength cryptographic digest algorithms (such as SHA-1) are not required, the use of lightweight custom digest algorithms with MDDT can substantially reduce computational overhead for MDDT client-server transactions.

- The technique can be implemented using any of a number of client-side and server-side programming languages and frameworks.

- MDDT can be used as a starting place for other techniques; the MDDT code and technique can be adapted for other applications.

- MDDT allows server-side access to information about users and their login sessions; thus, MDDT enables server-side software to control access, processing, and auditing to any desired level of granularity.

- MDDT can be made to work with more browsers and Web servers than HTTP digest authentication at this time.

- MDDT can be combined with encryption (for example, with HTTPS) to prevent eavesdropping.

- MDDT provides for clean session terminations through **logouts**.

- On balance, MDDT is a relatively lightweight solution that does not place significant computational burdens on clients or servers and does not require the services of external trusted servers or authorities.

MDDT suffers from these weaknesses:

- MDDT requires that every restricted Web page include a server-side script that verifies, prior to page loading, whether a request for the page comes from an authenticated user. While this is not a particularly burdensome requirement, it does add to the effort required to develop Web pages and it adds to the Web server's computational overhead.

- Use of the "back," "forward," or "refresh" buttons is discouraged so as to preserve the chain of authentication as users view access-restricted Web pages; this restriction is inconvenient not only for users, but for webmasters who must apply special HTML code for all links to other access-restricted pages on their site.

- To date, MDDT has not been tested for compatibility with any asynchronous JavaScript and XML (**AJAX**) page-design techniques.

- The original login window must be kept open during an MDDT login session because user credentials are stored in this window; this practice (as explained above in "Comparison of Resistance to Attacks") entails some risk. This requirement could be eliminated by modifying MDDT to store user credentials in **session cookies** (and optionally to transfer the ticket from server to client in session cookies); such a change might improve user convenience but would not improve security.

- As currently implemented, MDDT does not check the age of single-use tickets to insure that they are nullified if not used within a certain period of time, such as 15 or 20 minutes. Such a check can be added easily, and should be.

A Conjecture about Message-Dependent Digests

A message digest has value in that it represents a message without revealing information about the message; therefore, an algorithm for producing message digests should be, as nearly as possible, irreversible—there should not be any fast algorithm for identifying any part of a source message from its digest. When designing a message-digest algorithm, it would seem that the desirable trait of irreversibility is more likely to be achieved when the forward algorithm follows various paths of execution for different source messages, because then any algorithm for reversing the computation must test and traverse multiple reverse execution paths. These observations lead to the following informal conjecture: Relatively simple message-digest algorithms can be made irreversible for practical purposes by introducing algorithmic message dependency.

This conjecture is offered here for two purposes: First, to guide Web-application developers who modify MDDT with new message-digest functions, and second, to propose this as a topic for further investigation by cryptographic researchers (if there is not already a body of theory in this area). The truth or falsity of this conjecture is not critical to the utility of MDDT because the message-digest function used in MDDT can be replaced without changing overall functioning. Widely used message-digest algorithms, such as MD5 or SHA-1,[17] can be used in MDDT, although possibly at the cost of greater computational overhead.

Summary

MDDT (message-dependent digests with tickets) is a technique for restricting access to selected documents on an otherwise publicly accessible Web server, implemented using procedural code that runs under freely and widely available language translators: JavaScript on the client side and PHP on the server side. MDDT is well suited to Web sites and Web applications for which it is important to restrict access to interactive pages and operational forms. Like any other IT security control, MDDT should be applied or adapted carefully in consultation with IT-security specialists. If MDDT is used in a Federal application requiring high security, then the message-digest algorithm used must conform to *FIPS Publication 180–2* for the Secure Hash Standard (National Institute of Standards and Technology, 2002).

For some applications, other solutions may be more appropriate. In particular, HTTP Digest Authentication combined with PHP state management is similar to MDDT in capabilities and simpler to implement (Lerdorf and Tatroe, 2002). MDDT does, however, provide more tamper-resistant session continuity, superior server-side access to user identity and state (for fine-grained control of applications), and definite session logouts.

---

[17]MD5 and SHA-1 (Viega and others, 2002) both provide a feature that is not needed in MDDT; these digest algorithms produce almost no collisions (instances in which different source messages have the same digest). This feature is important when assurance of the integrity and authenticity of documents or digital files is required, but such absolute assurance is unnecessary in MDDT. If the probability of one collision is on the order of one in a million, the probability of two or three sequential collisions is the product of the probabilities for each collision; for practical purposes, this is a vanishingly small probability. Thus, a rare lucky guess by an attacker might yield access to a single Web page, but almost certainly not to any other pages.

References Cited

Apache HTTP Server Project, 2005, Authentication, authorization, and access control (digest caveat): Apache HTTP Server Project Web site at *http://httpd.apache.org/docs/1.3/howto/auth.html#digestcaveat*. (Accessed August 24, 2007.)

Apache HTTP Server Project, 2006, Apache module mod_auth_digest: Apache HTTP Server Project Web site at *http://httpd.apache.org/docs/2.2/mod/mod_auth_digest.html*. (Accessed August 24, 2007.)

Apache Week, 1996, Using user authentication: Apache Week Web site at *http://www.apacheweek.com/features/userauth/*. (Accessed August 24, 2007.)

Barrett, D.J., Silverman, R.E., and Byrnes, R.G., 2005, SSH: the Secure Shell (2d ed.): Sebastopol, Calif., O'Reilly & Associates, Inc., 645 p.

Bui, Sonny, and others, 2002, United States Patent No. US 6,412,007 B1—Mechanism for authorizing a data communication session between a client and a server: Google Patent Search Web site at *http://www.google.com/patents?vid=USPAT6412007*. (Accessed June 18, 2008.)

Garfinkel, Simson, Spafford, Gene, and Schwartz, Alan, 2003, Practical Unix & Internet security (3d ed.): Sebastopol, Calif., O'Reilly & Associates, Inc., 954 p.

IBM, 2006, Custom single login: IBM Web site at *http://publib.boulder.ibm.com/infocenter/tivihelp/v8r1/index.jsp?topic=/com.ibm.netcool_portal.doc/po20in/xF1310744573.html*. (Accessed August 24, 2007.)

Laurie, Ben, and Laurie, Peter, 1999, Apache—the definitive guide (2d ed.): Sebastopol, Calif., O'Reilly & Associates, Inc., 369 p.

Lerdorf, Rasmus, and Tatroe, Kevin, 2002, Programming PHP: Sebastopol, Calif., O'Reilly & Associates, Inc., 507 p.

National Institute of Standards and Technology, 2002, Federal Information Processing Standards Publication 180–2—Secure Hash Standard: National Institute of Standards and Technology (NIST) Web site at *http://csrc.nist.gov/publications/fips/fips180–2/fips180–2.pdf*. (Accessed June 18, 2008.)

Network Working Group, 1987, RFC 1034 - Domain names—concepts and facilities: Internet Engineering Task Force (IETF) Web site at *http://www.ietf.org/rfc/rfc1034.txt*. (Accessed September 7, 2007.)

Network Working Group, 1993, RFC 1518 - An architecture for IP address allocation with CIDR: Internet Engineering Task Force (IETF) Web site at *http://www.ietf.org/rfc/rfc1518.txt*. (Accessed September 7, 2007.)

Network Working Group, 1994, RFC 1738 - Uniform Resource Locators (URL): Internet Engineering Task Force (IETF) Web site at *http://www.ietf.org/rfc/rfc1738.txt*. (Accessed September 7, 2007.)

Network Working Group, 1998, RFC 2401 - Security architecture for the Internet Protocol: Internet Engineering Task Force (IETF) Web site at *http://www.ietf.org/rfc/rfc2401.txt*. (Accessed September 7, 2007.)

Network Working Group, 1999a, RFC 2246 - The TLS Protocol Version 1.0: Internet Engineering Task Force (IETF) Web site at *http://www.ietf.org/rfc/rfc2246.txt*. (Accessed September 7, 2007.)

Network Working Group, 1999b, RFC 2616 - Hypertext Transfer Protocol–HTTP/1.1: Internet Engineering Task Force (IETF) Web site at *http://www.ietf.org/rfc/rfc2616.txt*. (Accessed September 7, 2007.)

Network Working Group, 1999c, RFC 2617 - HTTP authentication—basic and digest access authentication: Internet Engineering Task Force (IETF) Web Site at *http://www.ietf.org/rfc/rfc2617.txt*. (Accessed September 7, 2007.)

Network Working Group, 1999d, RFC 2660 - The Secure HyperText Transfer Protocol: Internet Engineering Task Force (IETF) Web site at *http://www.ietf.org/rfc/rfc2660.txt*. (Accessed September 7, 2007.)

Network Working Group, 2000, RFC 2818 - HTTP over TLS: Internet Engineering Task Force (IETF) Web site at *http://www.ietf.org/rfc/rfc2818.txt*. (Accessed September 7, 2007.)

Network Working Group, 2003, RFC 3647 - Internet X.509 Public key infrastructure certificate policy and certification practices framework: Internet Engineering Task Force (IETF) Web site at *http://www.ietf.org/rfc/rfc3647.txt*. (Accessed September 7, 2007.)

Network Working Group, 2005, RFC 3986 - Uniform Resource Identifier (URI)–Generic Syntax: Internet Engineering Task Force (IETF) Web site at *http://www.ietf.org/rfc/rfc3986.txt*. (Accessed September 7, 2007.)

Open Web Application Security Project, 2002, A guide to building secure Web applications: cgisecurity.com Web site at *http://www.cgisecurity.com/owasp/html/*. (Accessed August 24, 2007.)

Open Web Application Security Project, 2007, Top 10 2007: Open Web Application Security Project (OWASP) Web site at *http://www.owasp.org/index.php/Top_10_2007*. (Accessed September 6, 2007.)

Pastore, Mike, and Dulaney, Emmett, 2004, Security +™ study guide (2d ed.): Alameda, Calif., SYBEX Inc., 500 p.

PHP Group, 2007, Chapter CXII. OpenSSL functions: PHP Group Web site at *http://www.php.net/manual/en/ref.openssl.php*. (Accessed August 24, 2007.)

Sasmazel, Levent MD, and Schneider, D.H., 2000, United States Patent No. 6,032,260—Method for issuing a new authenticated electronic ticket based on an expired authenticated ticket and distributed server architecture for using same: Google Patent Search Web site at *http://www.google.com/patents?vid=USPAT6032260*. (Accessed June 18, 2008.)

Skoudis, Ed, with Liston, Tom, 2006, Counter hack reloaded (2d ed.): Upper Saddle River, N.J., Prentice Hall, 748 p.

Stein, L.D., and Stewart, J.N., 2002, The World Wide Web security FAQ: World Wide Web Consortium (W3C) Web site at *http://www.w3.org/Security/Faq/*. (Accessed August 24, 2007.)

Venners, Bill, 2006, HTTP authentication woes: Artima Web site at *http://www.artima.com/weblogs/viewpost.jsp?thread=155252*. (Accessed August 24, 2007.)

Viega, John, Messier, Matt, and Chandra, Pravir, 2002, Network security with OpenSSL: Sebastopol, Calif., O'Reilly & Associates, Inc., 367 p.

World Wide Web Consortium, 2001, HTTP activity statement: World Wide Web Consortium Web site at *http://www.w3.org/Protocols/Activity.html*. (Accessed August 24, 2007.)

World Wide Web Consortium, 2008, World Wide Web Consortium (homepage): World Wide Web Consortium Web site at *http://www.w3.org/*. (Accessed January 17, 2008.)

Glossary

access control list   A list of identifying information used by computer software to control access to an information-system resource.

account harvesting   An attack consisting in the gathering of user names, passwords, or both for later attempts at unauthorized access to an information-system resource.

AJAX   Asynchronous JavaScript and XML. AJAX is an approach to the design of responsive, interactive Web pages.

algorithm   A step-by-step procedure for processing information or carrying out a computation.

attack   An activity directed towards gaining unauthorized access to data, information, or an information-system resource for any purpose (Skoudis with Liston, 2006).

authenticated   In reference to a user, having been duly identified as an authorized user (that is, as a person explicitly permitted to use an application or other information-system resource).

authentication   The state, process, or capability of identifying authorized users (persons permitted to use an application or other information-system resource).

authentication ticket   A complex and arbitrary character string or number used to grant temporary access to an information-system resource, such as a Web application.

authorized   Having been given permission to use an information-system resource.

client   The person, machine, or software that initiates requests for services or information during client-server interactions.

client-server   A two-way interaction between information-system resources in which a client issues requests for information and services and accepts responses, and a server responds to the client requests.

confidentiality   The ability to prevent access to an information-system resource by unauthorized users, or the quality of being protected from unauthorized disclosure or access.

cracking   Gaining unauthorized access to information by recovering the original message from an encrypted copy; unauthorized decryption.

credentials guessing   An attack consisting of repeated, usually automated, login attempts carried out in order to gain access.

cryptographic cipher   A procedure, algorithm, or process that reversibly converts a source file or message into an encoded form that is not directly usable.

data spoofing (tampering)   An attack consisting of unauthorized modification of data, either while the data are in transit over a network or residing on a Web server or other information system.

denial of service   An attack carried out to make an information-system resource temporarily or permanently unavailable through a concerted effort to overload the resource.

domain   A collection of information-system resources defined by IP address, the Domain Name System, or Microsoft® Active Directory®.

Domain Name System   A collection of Internet servers and services that associate and cross-reference human-readable names with IP addresses (Network Working Group, RFC 1034, 1987).

eavesdropping   The process or activity of secretly obtaining information transmitted over a network between two other parties without stopping or detectably slowing the transmission.

encryption   The process or activity of reversibly encoding information into a form that is not directly usable or understandable.

extranet   A network or network-based resource available to users on an internal network (an intranet) and to a limited external clientele.

hash   The process or result of reproducibly transforming any ordered collection of bytes of any length into a fixed-length string or number (National Institute of Standards and Technology, 2002). Some hashes are used as message digests.

HTTPS   HyperText Transfer Protocol Secure specifies HTTP over Secure Sockets Layer, TLS, or OpenSSL (Network Working Group, RFC 2818, 2000).

HyperText Transfer Protocol   A lightweight protocol for transferring files via the Internet; the defining protocol for the World Wide Web (Network Working Group, RFC 2616, 1999b).

include file   A file referenced in another file with an "include" directive. (The "include" directive within a primary file references a secondary include file and directs that the include file's contents be retrieved and incorporated into the contents of the primary file.)

information technology   Hardware, software, and methods of processing and communicating information; especially computers, communications networks, and other digital electronic devices used for processing information.

integrity   The state of being complete and free from unauthorized modifications.

interception   The process or act of obtaining information intended for another recipient in such as way as to prevent its receipt by the intended recipient.

internet   The global network of networks.

Internet service provider   A firm or other organization that provides equipment and services that enable networks or users to connect to the Internet.

intranet   An internal or private network that may or may not be connected to the Internet.

IP address   An address for an hardware network interface as defined by the Internetworking Protocol (IP) (Network Working Group, 1993, RFC 1518). (Under IPv4, an IP address consists of four octets of bits, for a total of 32 bits.)

IP spoofing   Communicating on a network with one or more forged IP addresses.

IPsec   A protocol widely used for creating Virtual Private Networks by encrypting communications between a publicly accessible network and a private network (Network Working Group, RFC 2401, 1998).

login   The act or process by which a user is identified (authenticated) to an information-system resource as someone authorized to access the resource. (Typically, a login involves an exchange that demonstrates that the user possesses valid credentials, such as a user name and password, though the credentials themselves may not necessarily be exchanged.)

login session   A finite period of time following the completion of a login during which a user is known to an information-system resource as an authorized user and allowed to make use of the resource.

logout   In reference to a Web application, a process that ends a login session unequivocally by removing or setting server-side state variables to close the session and by sending a Web page to the user reporting the session termination.

man-in-the-middle   person-in-the-middle.

MDDT   The technique for secure (that is, restricted) Web-site access described in this publication: message-dependent digests with tickets.

message-dependent   In reference to a message digest, created with an algorithm that prescribes different processing steps, depending on the content of the message to be digested. (Alternatively, a message-dependent digest may be said to be a digest created by one member of a parametric family of algorithms, with the elements of the source message providing the parameters by which the member algorithm is determined.)

message digest   A compact string or a number reproducibly derived from any collection of bytes, such as a character string, a digital document, or any other computer file. (For practical purposes, a digest should be, as measured in bytes of information, much shorter than the source message. Typically message digests are no more than 30 or 40 bytes.)

nonrepudiation   In reference to a transaction or exchange, a quality of verifiability such that no party to the transaction or exchange may credibly repudiate or deny occurrence and details of the transaction or exchange.

nonce   An arbitrary, semantically void, and highly improbable character string, number, or value generated for a single use ("for the nonce").

OpenSSL   "… an open-source library that implements the SSL and TLS protocols, and is the most widely deployed, freely available implementation of these protocols" (Viega and others, 2002).

packet sniffer   Software or a combination of software and hardware that eavesdrops on network transmissions.

person-in-the-middle   An attack (also known as a man-in-the-middle attack) in which a person secretly monitors and alters transmissions between two other parties. (The person-in-the-middle, or PITM, attack typically involves a proxy that may operate on either the client's network or the server's network.)

proxy   A server, workstation, or other information-system resource that mediates client-server transactions. (Some proxies are authorized and legitimate while others operate secretly and without authorization by network administrators.)

public key infrastructure   The collection of widely available software, procedures, legal agreements, and certification authorities that establish trust among those who use public-key cryptography to secure transactions and verify digital signatures. The topics of public-key cryptography and digital signatures are not discussed in detail in this publication due to their complexity (Viega and others, 2002; Network Working Group, RFC 3647, 2003).

replay   An attack in which captured credentials or session keys are used later to gain unauthorized access to an information-system resource.

Secure HyperText Transfer Protocol   A protocol (S-HTTP) for secure transmission of messages using HTTP (Network Working Group, RFC 2660, 1999d).

Secure Sockets Layer   A protocol based on public-key cryptography for transmission of encrypted information over the Internet.

server   Hardware or software that responds to requests for information or services initiated by clients.

session   A finite period of time during which a user interacts with an information-system resource.

session cloning (client-side masquerading)   An attack in which an unauthorized person obtains session-related variables and uses them to take over a user session as the client.

session continuity   The quality of unbroken access to, and availability of, an information-system resource during a session.

session cookie   An item of state data stored by a Web browser on the client workstation only for the duration of the Web-browser session. Session cookies are deleted when a Web-browser window is closed or, at the latest, when the Web browser is shut down.

session hijacking (server-side masquerading)   An attack in which a client's interactions with a chosen server are redirected to another server without the client's permission or awareness.

social engineering   The use of deception or manipulation in social interchanges to obtain information useful for gaining unauthorized access to information-system resources.

state   The condition or mode of a transaction or process.

state management   The processes and facilities by which an interactive information-systems resource keeps track of its own state and that of all of its users.

stateless   In reference to a protocol, having no history by which a current transaction can be affected by a prior transaction.

subnet   A portion of the addresses or devices on a network. (Usually a subnet is defined by the portion of an address or naming string that can be modified in order to generate addresses or names within the subnet.)

ticket   A unique or highly improbable string, document, number, or value generated for granting temporary access to an information-system resource; a kind of nonce.

Transport Layer Security   The protocol that is the successor to Secure Sockets Layer (Network Working Group, RFC 2246, 1999a) for transmission of encrypted information over the Internet.

Uniform Resource Locator   Loosely, a Uniform Resource Identifier (URI), that is, a relatively compact name for a document or other resource accessible through the Internet (Network Working Group, RFC 1738, 1994; Network Working Group, RFC 3986, 2005).

URL guessing   An attack performed by guessing a URL in order to view a Web document or other Internet resource for which the URL is unpublished or otherwise restricted.

virtual host   A Web site hosted on a server configured to host multiple sites, each with its own individual domain name; or a server configured for hosting multiple, independently named Web sites.

Virtual Private Network   A set of facilities by which users may securely access a private network through a public network.

Web   A short name for the World Wide Web.

Web application   A collection of software for performing interactive processing through Web forms and Web pages; a computer application with a Web browser as the user interface.

Web client   A visitor to a Web site, a user of a Web application, or the Web browser used by a Web-site visitor or Web-application user.

Web server   The hardware, software, or both, that respond(s) to HTTP client requests.

webmaster   An individual or team that operates a Web site.

Web site   A specific named collection of documents and forms available for HTTP access on the Internet or on a private network.

workaround   A response to a problem that avoids the problem (by working around it) rather than solving it.

World Wide Web   The total, world-wide collection of Internet-accessible Web sites and their collective content (World Wide Web Consortium, 2008).

World Wide Web site   A Web site. ("Web site" is the more common usage.)

WWW   The World Wide Web.

Appendixes 1–2

The following appendixes provide (1) a list of abbreviations used throughout the text of this publication and (2) working JavaScript and PHP code for an implementation of MDDT.

Appendix 1.   List of Abbreviations

ACL	access control list
AJAX	asynchronous JavaScript and XML
DoS	denial of service
DNS	Domain Name System
HTTP	HyperText Transfer Protocol
HTTPS	HyperText Transfer Protocol Secure
IP	Internetworking Protocol
ISP	Internet service provider
IT	information technology
MITM	man-in-the-middle
PHP	PHP HyperText Preprocessor
PITM	person-in-the-middle
PKI	public key infrastructure
RFC	Request for Comments
S-HTTP	Secure HyperText Transfer Protocol
SSL	Secure Sockets Layer
TLS	Transport Layer Security
URL	Uniform Resource Locator
VPN	Virtual Private Network
WWW	The World Wide Web

Appendix 2.   Working JavaScript and PHP Code

This appendix contains JavaScript and PHP code from a working implementation of MDDT.[1] Although these programs have been used by the USGS, no warranty, expressed or implied, is made by the USGS or the United States Government as to the accuracy and functioning of these programs and related program material nor shall the fact of distribution constitute any such warranty, and no responsibility is assumed by the USGS in connection therewith. The working code consists of:

- enough PHP pages (four) to illustrate how the PHP and JavaScript include files[2] are referenced from login and application pages,

- a list of code excerpts illustrating how to link and pass parameters among application pages,

- five PHP include files (code incorporated into PHP pages with the "include" directive), and

- two JavaScript include files (code incorporated into the scripts on PHP pages with the "include" directive).

The code files in this Appendix contain several long URLs or PHP identifiers broken across lines because of the limited page width. To use these files as working code, these URLs and identifiers must be rejoined.

PHP Pages

index.php (Login.php)   First page seen during authentication

AppAuthenticate.php   Unseen authentication-check page

AppHome.php   Homepage for successfully authenticated users

/management/index.php   Sample page for demonstrating secure links

Code excerpts   Excerpts illustrating various secure link types

In the listings for these PHP pages, the portion of page code that is specific to MDDT access restriction is in bold-face blue type for emphasis.

PHP Include Files

MakeTicket.phpinc   Routine for creating a ticket

SaveTicket.phpinc   Routine for saving the ticket to a file

HiddenForm01.phpinc   Page element for supplying the ticket in a hidden variable

InitAuth.phpinc   Server-side code for initial authentication

ContAuth.phpinc   Server-side code for continuing use of restricted pages

JavaScript Include Files

InitLogin.js   Client-side code for initial authentication

ContLogin.js   Client-side code for continuing use of restricted pages

---

[1]The MDDT code in this appendix is from the implementation for the USGS Eastern Geographic Science Center (EGSC) High-Performance Computing Cluster (HPCC) Operational Web Site (*http://egscbeowulf.er.usgs.gov/*).

[2]An **include file** is a file that is incorporated into a referring page or script during processing. The include file is referenced in the referring (or "calling") page or script by an "include" directive.

PHP Pages

index.php (aka Login.php): First page seen during authentication

```
<html>
<head>
    <title>EGSC HPCC Applications Section Main Page: Authentication for Application
    Access</title>
    <link rel="stylesheet" type="text/css" href="/css/hpcc01a.css">
    <script language="JavaScript">
    <? include '/var/www/html/applications/php/MakeTicket.phpinc'; ?>
    <? include '/var/www/html/applications/php/SaveTicket.phpinc'; ?>
    <? include '/var/www/html/applications/js/InitLogin.js' ?>
    </script>
</head>
<body bgcolor="#f0f0f0">

<? include '/var/www/html/includes/header2c.html'?>

<table width="840">
    <tr>
        <td class="FormalColumnT" colspan="2" height="60"><h2>Authentication
        for Application Access</h2></td>
    </tr>
    <tr>
        <td class="FormalColumnL" width="200" height="250">Use of applications
        on the USGS Eastern Geographic Science Center (EGSC) High-Performance
        Computing Cluster(HPCC) is restricted to authorized users. If you have
        not been provided with login credentials, please
        <a href="http://egscbeowulf.er.usgs.gov/">exit</a> from this page.
        Information about becoming an authorized user is
        <a href="http://egscbeowulf.er.usgs.gov/AuthUserInfo.php"> available on
        the public area </a> of this Web site.

        <td class="FormColumn" width="70%" height="250">
        <form name="form1" action="AppAuthenticate.php" target="_blank"
        onsubmit="SaveAUvalues(this);">
        <center>
            <table>
                <tr>
                    <td class="FormItemDescription">User Identification (UserID):</td>
                    <td><input type="text"      name="aus1" maxlength="50"></td>
                </tr>
                <tr>
                    <td class="FormItemDescription">Authorization Code:</td>
                    <td><input type="password" name="aus2" maxlength="50"></td>
                </tr>
            </table>
<br/><input type="submit" name="aus3" value="Enter Access Information"><br/>
        </center>
        <input type="hidden" name="hpccTicket" value="<? printf("%s",$Ticket); ?>">
        </form>
<br/><br/>
```

```
<p><b>Please Note:</b>
    <ul>
        <li>Logging in automatically opens a new window for your HPCC work.
        While you use HPCC applications in the new window, this window
        must remain open displaying the Authorized-Users Login Page. This
        window must not be used again until after logout, but you may
        minimize it if you wish.</li>

        <li>You must refresh this page between logins. With some browsers
        (including Mozilla Firefox) you must leave the page and return. With other
        browsers (including Internet Explorer 6.0) it is only necessary to click the
        "Refresh" button.</li>

        <li>Please do not use your browser's "Back", "Forward", or "Refresh"
        buttons during your HPCC login session.</li>

        <li>Please do not allow your browser to save ("remember") your login
        credentials.</li>

        <li>Please do not attempt to open more than one HPCC login session
        from the same workstation. Only one login session from a single
        IP address is allowed at any one time.</li>
    </ul>
</p>

        </td>
    </tr>
</table>

<? include '/var/www/html/includes/footer2.html' ?>

</body>
</html>
```

AppAuthenticate.php: Unseen authentication-check page

```
<?
    // This PHP code tests for valid userid and password and re-directs either
    // to the authenticated-users homepage or a failure page.
    // If the user is sent to the AU homepage, a new ticket and digest are
    // generated so the AU homepage will be able to perform the ticket-based
    // authentication used by all actual pages in the secure are of the Web site.
    include '/var/www/html/applications/php/InitAuth.phpinc'; $val = InitAuth();
    if (substr($val,0,3) == "Bad") {header('Location: http://egscbeowulf.er.usgs.gov/
    LoginFailed.php'); exit();}
    include '/var/www/html/applications/php/MakeTicket.phpinc';
    include '/var/www/html/applications/php/SaveTicket.phpinc';
    $NewDigest = Digest01a($Ticket, $uid, $pw);
    $locstring = "Location: http://egscbeowulf.er.usgs.gov/applications/
    AppHome.php?hpccTicket=";
    $locstring = $locstring.$NewDigest;
    header($locstring); exit();
?>

<html>
<head>
    <title>EGSC HPCC Applications: User-Authentication Processing</title>
    <link rel="stylesheet" type="text/css" href="/css/hpcc01a.css">
</head>
<body>

<? include '/var/www/html/includes/header2a.html'?>

<br/><br/>
<table width="800">

    <tr>
        <td class="FormalColumn" colspan="2" height="60px"><h2>User
        Authentication</h2></td>
    </tr>
    <tr>
        <td class="FormalColumn" height="250px">
<p class="FeaturedItem">The HPCC authentication system has malfunctioned.
Please report the contents of this page to the
<a href="mailto:didonato@usgs.gov">EGSC HPCC manager</a>.
        </td>
    </tr>

</table>
<br/><br/>

<? include '/var/www/html/includes/footer2.html' ?>

</body>
</html>
```

AppHome.php: Homepage for successfully authenticated users

```php
<?
    include '/var/www/html/applications/php/ContAuth.phpinc'; $val = ContAuth();
    if ($val == "Bad") {header('Location: http://egscbeowulf.er.usgs.gov/
    AuthenticationFailure.php'); exit();}
    include '/var/www/html/applications/php/MakeTicket.phpinc';
    include '/var/www/html/applications/php/SaveTicket.phpinc';
?>
<html>
<head>
    <title>EGSC HPCC Applications: Homepage</title>
    <link rel="stylesheet" type="text/css" href="/css/hpcc01a.css">
    <script language="JavaScript">
    <? include '/var/www/html/applications/js/ContLogin.js' ?>
    </script>
</head>
<body>


<? include '/var/www/html/php/SubHeader1.phpinc'; SubHeader1("EGSC HPCC Operations",
"Applications", "SubHeadc"); ?>

<table width="840">

    <tr>
        <td class="FormalColumnT" colspan="2" height="60">
        <h2>EGSC HPCC Applications Homepage</h2>
<p>All applications available on the EGSC HPCC are shown and may be accessed
from this Applications Homepage, though
during this session you may only use the applications for which <span class="green"><?
print($uid); ?></span> is a valid user identifier.
</p>
        </td>
    </tr>
    <tr>
        <td class="MenuColumn" width="200" height="250">

        <h3>Selections<h3>
        <table valign="center" align="center">

            <tr>
                <td class="MenuItem">
                <a class="Menu" onclick="LinkOut(this)" href="/applications/compstat/
                index.php">Computational Statistics</a>
                </td>
            </tr>

            <tr>
                <td class="MenuItem">
                <a class="Menu" onclick="LinkOut(this)" href="/applications/geonames/
                index.php">Geographic Names Rapid Search</a>
                </td>
            </tr>
```

```
    <tr>
        <td class="MenuItem">
        <a class="Menu" onclick="LinkOut(this)" href="/applications/
        management/
        index.php">HPCC Management and System Administration</a>
        </td>
    </tr>

    <tr>
        <td class="MenuItem">
        <a class="Menu" onclick="LinkOut(this)" href="/applications/jpeg2000/
        index.php">JPEG2000 Encoding</a>
        </td>
    </tr>

    <tr>
        <td class="MenuItem">
        <a class="Menu" onclick="LinkOut(this)" href="/applications/weather/
        index.php">Meteorological Data Selection</a>
        </td>
    </tr>

    <tr>
        <td class="MenuItem">
        <a class="Menu" onclick="LinkOut(this)" href="/applications/nmmft/
        index.php">National Model of Mercury in Fish Tissue
        (NMMFT) Calibration</a>
        </td>
    </tr>

    <tr>
        <td class="MenuItem">
        <a class="Menu" onclick="LinkOut(this)" href="/applications/sleuth/
        index.php">SLEUTH (Urban Growth) Model</a>
        </td>
    </tr>

    <tr>
        <td class="MenuItem">
        <a class="Menu" onclick="LinkOut(this)" href="/applications/utilities/
        index.php">User Utilities</a>
        </td>
    </tr>
    </table>

</td>
<td class="FormalColumnR" width="70%" height="250">
<h4>Applications on the EGSC HPCC</h4>

<p>The menu on the left will take you to an application or an application set.
Selecting a menu item <i>does not</i> initiate the application &#151; this just
takes you to the application's main page from which you may perform the operations
available for that particular application. Each application is responsible for
insuring its own user restrictions. If you have not been authorized to use an
application, selecting it from the menu on the left will send you to an error page.
```

Please remember not to use the "Back", "Forward", or "Refresh" buttons on your browser
while using applications since using these buttons breaks your user-authentication
chain and returns you to an error page.
</p>

```
            <table style="border-color:red; border-width: 4px; border-style:
            solid; padding: 8px;" >
                <tr>
                    <td>
                    <h3 class="Red">IMPORTANT NOTICE TO USERS OF THIS SYSTEM</h3>
                    <p class="Notice2">
                    This is a United States Government computer system, maintained by
                    the Department of the Interior, to provide official, unclassified
                    U.S. Government information. Use of this system by any user,
                    whether authorized or unauthorized, constitutes consent to
                    monitoring, recording,
                    and disclosure of use by authorized personnel.
                    <b>Users have no reasonable expectation of privacy in the use
                    of this system.</b>

                    Unauthorized use may subject violators to criminal, civil,
                    or disciplinary action.
                    </p>
                    </td>
                </tr>
            </table>
        </td>
        <td>
        <? include '/var/www/html/applications/php/HiddenForm01.phpinc' ?>
        </td>
    </tr>

</table>

<? include '/var/www/html/applications/php/BottomNavMain1.phpinc'; BottomNavMain1();
?>
<? include '/var/www/html/applications/php/FooterMain1.phpinc'; FooterMain1(); ?>

</body>
</html>
```

/management/index.php: Sample page for demonstrating secure links

```
<?
    include '/var/www/html/applications/php/ContAuth.phpinc'; $val = ContAuth();
    if ($val == "Bad") {header('Location: http://egscbeowulf.er.usgs.gov/
    AuthenticationFailure.php');}
    include '/var/www/html/applications/php/MakeTicket.phpinc';
    include '/var/www/html/applications/php/SaveTicket.phpinc';
    include '/var/www/html/applications/php/AppAuth.phpinc'; $val =
    AppAuth("Management");
    if ($val == "Bad")
        {
        $NewDigest = Digest01a($Ticket, $uid, $pw);
        $locstring = "Location: http://egscbeowulf.er.usgs.gov/applications/
        AppAuthenticationFailure.php?hpccTicket=";
        $locstring = $locstring.$NewDigest;
        header($locstring); exit();
        }
    include '/var/www/html/applications/php/TaskProg.phpinc';
    TaskProg("Management Application Home", "Entered homepage for the EGSC HPCC
    Management application.");
?>
<html>
<head>
    <title>EGSC HPCC Applications: EGSC HPCC Management</title>
    <link rel="stylesheet" type="text/css" href="/css/hpcc01a.css">
    <script language="JavaScript">
    <? include '/var/www/html/applications/js/ContLogin.js' ?>
    </script>
</head>
<body>

<? include '/var/www/html/php/SubHeader1.phpinc'; SubHeader1("EGSC HPCC Operations",
"Applications", "SubHeadc"); ?>
<? include '/var/www/html/applications/php/AppHeader1.phpinc'; AppHeader1("HPCC
Management"); ?>
<? include '/var/www/html/applications/php/TopNavMain1.phpinc'; TopNavMain1("HPCC
Management", "/applications/management/index.php"); ?>

<table width="840">
    <tr>
        <td class="FormalColumnT" colspan="2" height="40">
        <h2>EGSC HPCC Management</h2>
        </td>
    </tr>
    <tr>
        <td class="MenuColumn" width="200" height="250">

        <h3>Selections<h3>
            <table valign="center" align="center">

                <tr>
                    <td class="MenuItem">
                    <a class="Menu" onclick="LinkOut(this)" href="/applications/
                    management/backups.php">Backups</a>
                    </td>
                </tr>
```

```
<tr>
    <td class="MenuItem">
    <a class="Menu" onclick="LinkOut(this)" href="/applications/
    management/cleanup.php">Clean Up Backup Files</a>
    </td>
/tr>

<tr>
    <td class="MenuItem">
    <a class="Menu" onclick="LinkOut(this)" href="/applications/
    management/status.php">Cluster Status</a>
    </td>
</tr>

<tr>
    <td class="MenuItem">
    <a class="Menu" onclick="LinkOut(this)" href="/applications/
    management/annlogcm.php">Enter HPCC Management Notes</a>
    </td>
</tr>

<tr>
    <td class="MenuItem">
    <a class="Menu" onclick="LinkOut(this)" href="/applications/
    management/annlogcm.php">Log and Manage Change</a>
    </td>
</tr>

<tr>
    <td class="MenuItem">
    <a class="Menu" onclick="LinkOut(this)" href="/applications/
    management/miscoprs.php">Miscellaneous Administrative Operations</a>
    </td>
</tr>

<tr>
    <td class="MenuItem">
    <a class="Menu" onclick="LinkOut(this)" href="/applications/
    management/shutdown.php">Shut Down Work Nodes</a>
    </td>
</tr>

<tr>
    <td class="MenuItem">
    <a class="Menu" onclick="LinkOut(this)" href="/applications/
    management/viewmenu.php">View User and Management Entries</a>
    </td>
</tr>

<tr>
    <td class="MenuItem">
    <a class="Menu" onclick="LinkOut(this)" href="/applications/
    management/checklist.php">Weekly Administrative Checklist</a>
    </td>
</tr>
```

```
                </table>
            </td>
            <td class="FormalColumnR" width="70%" height="250">
            <h4>Managing the EGSC HPCC</h4>
            <p>This application set provides a number of operations useful for
            managing the EGSC HPCC.
            Some of the facilities of this set will be available to application users
            through the Utilities application set. This set, however, is for the
            exclusive use of
            the EGSC HPCC manager and System Administrator.
            </p>
            <? include '/var/www/html/applications/php/HiddenForm01.phpinc' ?>
            </td>
        </tr>

</table>

<? include '/var/www/html/applications/php/BottomNavOther1.phpinc';
BottomNavOther1("HPCC Management", "/applications/management"); ?>
<? include '/var/www/html/applications/php/FooterOther1.phpinc'; FooterOther1(); ?>

</body>
</html>
```

Code excerpts: Excerpts illustrating various secure link types

LinkOutF : Link from a form

```
<form name="alcmform1" action="annlogcmFP.php" method="post"
onsubmit="LinkOutF(this)">
<br/>
Type of Entry:<br/>
<input type="hidden" name="hpccTicket" value="<? print($Ticket); ?>">
<input type="radio" name="enttype" value="Administrative Note">       Administrative
Note<br/>
<input type="radio" name="enttype" value="Change Annotation">         Change
Annotation<br/>
<input type="radio" name="enttype" value="Miscellaneous Note" checked> Miscellaneous
Note<br/>
<input type="radio" name="enttype" value="Software-Development Note">  Software-
Development Note<br/><br/>
Keywords for this Entry:<br/>
<input    name="keywords" type="text" size="40" maxlength="80" value="  "><br/><br/>
Your Name:<br/>
<input    name="pc"       type="text" size="40" maxlength="80" value="  "><br/><br/>
Your Note:<br/>
<textarea name="detailinfo" rows="10" cols="50" wrap="virtual" > </
textarea><br/><br/><br/>

<center><input type="submit" value="Submit Your Note"></center>

</form>
```

LinkOutP : Link from an anchor and pass a parameter

```
    $tv1 = intval( (time() - $tmax[3])/86400 );
    print("<tr><td $td1>[ 3] Review SSH Log</td><td align=\"center\" $td1>$tv1</td>");
    print("<td $td1>$tdstring[3]</td>");
    print("<td align=\"center\" $td1><a href=\"checklistFP.php\" onclick=\"LinkOutP(t
his,'&opcode=3')\">
        Perform</a></td></tr>\n");
```

LinkOutRefreshP : Let a page that takes parameters refresh itself from a button or anchor

```
$Fname=raw_param('fname'); $Fsize=raw_param('fsize');
if ($Fsize > $tarr8[4])
{
  print("<p>Before the file $Fname can be downloaded, it must be copied to a directory
within \n");
  print("the file-system branches directly accessible to the Web server. The file
contains $Fsize
        bytes.\n");
  print("Copying generally requires about a minute per gigabyte. <span
class=\"red\">You must periodically
        \n");
    print("press the \"/Refresh Status\" button until copying is complete and
```

```
      $Fname is ready to be downloaded.</span></p>\n");
      print("<center><input type=\"button\" value=\"Refresh Status\"
      action=\"bkdown.php\"
      onclick=\"LinkOutRefreshP(0)\"></center>");
}


foreach ($rarray2 as $v1)
{
      $t4   = number_format($tarr8[4]);
      print("<tr><td>$tarr8[5] $tarr8[6] $tarr8[7]</td>
      <td align=\"right\">$t4</td>");
      /* The following link allows downloading of a gzipped file by right clicking. */
      print("<td><a onclick=\"LinkOutRefreshP(0)\" type=\"application/x-gzip\"
      href=\"/applications/management/backups/$Fname\">$tarr8[8]</a></td></tr>");
}
```

LinkOutU: Relay a page's incoming parameters to a new page.

```
<?
    include '/var/www/html/applications/php/ContAuth.phpinc'; $val = ContAuth();
    if ($val == "Bad") {header('Location: http://egscbeowulf.er.usgs.gov/
    AuthenticationFailure.php');}
    include '/var/www/html/applications/php/MakeTicket.phpinc';
    include '/var/www/html/applications/php/SaveTicket.phpinc';
    include '/var/www/html/applications/php/AppAuth.phpinc';
    $val = AppAuth("Management");
    if ($val == "Bad")
    {
    $NewDigest = Digest01a($Ticket, $uid, $pw);
    $locstring = "Location: http://egscbeowulf.er.usgs.gov/applications/
    AppAuthenticationFailure.php?hpccTicket=";
    $locstring = $locstring.$NewDigest;
    header($locstring); exit();
    }
    $Fname=raw_param('fname'); $Fsize=raw_param('fsize');
    $fsresult = filesize("/data/www/backups/".$Fname);
    if (!$fsresult || $fsresult > 2000000000)
    {
    exec("/data/www/setuids/root/bsplitter02 $Fname > /dev/null&");
    }
    else
    {
    exec("cp /data/www/backups/$Fname /var/www/html/applications/management/
    backups/$Fname>/dev/null&");
    }
?>
<html>
<head>
    <title>EGSC HPCC Applications: EGSC HPCC Management - Backup-File Copy</title>
    <link rel="stylesheet" type="text/css" href="/css/hpcc01a.css">
    <script language="JavaScript">
    <? include '/var/www/html/applications/js/ContLogin.js' ?>
    </script>
</head>
<body onload="LinkOutU('/applications/management/bkdown.php');">
/* The contents of the body of this page won't ever be seen unless there is an error
in the processing which precedes generation of the page. */
</body>
</html>
```

PHP Include Files

MakeTicket.phpinc : Routine for creating a ticket

```
<?
/*************************************************************


    Function Name:                  MakeTicket
    File Name:                      MakeTicket.phpinc
    System:                         EGSC HPCC
    Date Created:                   Jan. 19, 2005
    Modification Dates:             N/A


    Purpose and Use:

    This function creates a ticket from the date, time, and
    client's IP address. A "ticket", as the term is used here,
    is an ordered set of bytes of finite length. The ticket
    is intended to be distinctive and difficult to
    predict. The ticket created by this function will be
    passed to Web clients to be used in encrypting or
    creating digests of authentication strings so that
    each communication between client and server will send
    essentially unpredictable information even during the
    same user session. (This inhibits unauthorized access
    by use of packet sniffing.) The ticket contains only
    printable characters (guaranteed by the helper
    function "modify").

    The function takes a single argument containing
    the client's IP address as a string as its single
    argument.

    Modification Notes:

*************************************************************/

function MakeTicket($IPstring)
{
    $format1 = "l, F d, Y :: h:i:s a";
    $currenttime = time();
    $wstr1 = date($format1, $currenttime);
    srand(time());
    $shufflestring = "";
    for ($j=0; $j<50; $j++)
        {
        $c1 = chr(rand(33,126));
        $shufflestring = $shufflestring.$c1;
        }
//  $shufflestring = "9#wL;5jJnqzP[c_=/eDVxx793^&/'ZY4*8@]}{?&r!+iM";
    $IPstring = shuffler($IPstring, $shufflestring);
    $wstr2 = shuffler($shufflestring, $IPstring);
    $wstr3 = modify($wstr1, $wstr2);
    $wstr4 = shuffler($wstr3, $wstr2);
    return $wstr4;
}
```

```
function shuffler($tobeshuffled, $shufflestring)
{
    $len1 = strlen($tobeshuffled);
    $index1 = 0;
    $newstring = $tobeshuffled;

    for ($i=0; $i<$len1; $i++)
        {
        $v1 = ord( substr($shufflestring,$index1++,1) );
        $pos = $v1%$len1;
        $temp = $tobeshuffled[$i];
        $c1 = $tobeshuffled[$pos];
        switch($i)
        {
            case 0:
            $tobeshuffled = $c1.substr($tobeshuffled, 1);
            break;
            case $len1-1:
            $tobeshuffled = substr($tobeshuffled, 0, $len-1).$c1;
            break;
            default:
            $tobeshuffled = substr($tobeshuffled, 0, $i).$c1.substr($tobeshuffled, $i+1);
            break;
        }
        switch($pos)
        {
            case 0:
            $tobeshuffled = $temp.substr($tobeshuffled, 1);
            break;
            case $len-1:
            $tobeshuffled = substr($tobeshuffled, 0, $len-1).$temp;
            break;
            default:
            $tobeshuffled = substr($tobeshuffled, 0, $pos).$temp.substr
            ($tobeshuffled, $pos+1);
            break;
        }
            if ($index1 >= strlen($shufflestring)) $index1 = 0;
        }
    return $tobeshuffled;
    //return $debugstr;
}

function modify($tobemodified, $modstring)
{
    $len1 = strlen($tobemodified);
    $index1 = 0;
    for ($i=0; $i<$len1; $i++)
        {
        $v1 = ord( substr($tobemodified, $i, 1) );
        $v2 = ord( substr($modstring, $index1++, 1) );
        $v3 = ($v1 + $v2)%127;
            if ($v3 < 32) {$v3 = $v3 + 33;}
            if ($v3 == 32){$v3 = $v3 + rand(1,20);}
            if ($v3 == 34){$v3 = rand(35,126);}
```

```
        if ($v3 == 96){$v3 = 126;}
    $c3 = chr($v3);
    $tobemodified = substr_replace($tobemodified, $c3, $i, 1);
    if ($index1 >= strlen($modstring)) $index1 = 0;
    }
    return $tobemodified;
}
$Ticket = MakeTicket($_SERVER['REMOTE_ADDR']);
?>
```

<u>SaveTicket.phpinc : Routine for saving the ticket to a file</u>
```
<?
/**********************************************************


        Function:               SaveTicket
        File Name:              SaveTicket.phpinc
        System:                 EGSC HPCC
        Author:                 David I. Donato
        Date:                   January 21, 2005
        Modification Dates:     August 31, 2005
                                September 2, 2005


        Purpose and Use:

        This function writes the ticket sent to a client
        in a file unique to the IP address of the client.
        The ticket written to the file will be retrieved
        and used to decode submissions from the client
        in such a way as to insure that the submissions
        have come from a client in possession of the
        ticket. This function is part of the collection
        of functions used to maintain security and to
        restrict use of certain pages to authorized users.

        The function takes the ticket as its single argument.

        Modifications:

            August 31, 2005 -- A system() call to 'sync' was
            reluctantly added because of authentication
            failures that seem to have no explanation
            other than the reading of an old ticket
            rather than the new one, which is waiting to be
            written to disk. Although the OS should prevent
            such situations, I hypothesize that it occurs
            because of Apache's use of multiple processes
            for servicing Web requests.

            September 2, 2005 -- The use of the system call to
            'sync' creates a problem: if a page spawns a
            background process which writes to a large file,
            then no new page can be accessed until the write
            operation has completed because the new page's
            SaveTicket operation will have to wait for 'sync'
            to complete before returning to the page to let
            it complete page generation. The system call to
            'sync' was replaced by an "fflush(fd);" statement.
            I hope this will resolve this issue.


***********************************************************/

function SaveTicket($tokstr)
{

    $IPaddr = $_SERVER['REMOTE_ADDR'];
    $filename = "/data/www/uauth/specific/CL".$IPaddr.".dat";
```

```
    $F01 = fopen($filename, "w");
    fwrite($F01, $tokstr);
    fflush($F01);
    fclose($F01);
    /***  system("sync"); ***/

}

SaveTicket($Ticket);
?>
```

HiddenForm01.phpinc: Page element for supplying the ticket in a hidden variable

```
<?
/*********************************************************


    Function:              Hidden Form # 01
    File Name:             HiddenForm01.phpinc
    System:                EGSC HPCC
    Date:                  January 26, 2005
    Modification Dates:    N/A


    Purpose and Use:

    This function creates a form containing a single
    hidden variable to hold the ticket. The form is
    not visible in the browser window. The function
    must receive the ticket as its single argument.


    Modifications:


*********************************************************/

function HiddenForm01($ticket)
{

    printf("<form name=\"TicketForm\"><input type=\"hidden\" name=\"hpccTicket\"
    value=\"");
    printf("%s", $ticket);
    printf("\"></form>");
    return;
}


HiddenForm01($Ticket);
?>
```

__InitAuth.phpinc: Server-side code for initial authentication__

```
<?
/***********************************************************

    Function:                 InitAuth
    File Name:                InitAuth.phpinc
    System:                   EGSC HPCC
    Date:                     January 24, 2005
    Modification Dates:       N/A

    Purpose and Use:

    This function carries out the initial authentication
    of a user logging in to this site. This function
    processes as follows:

        (1) Reads in the server-side ticket for the client IP
        address from the file;
        (2) Loops through possible UserID-Password pairs
        attempting to reproduce the client-side ticket
        passed to this script;
        (3) If a match is found, stores the UserID and password
        in a new file identified by the IP address and
        creates a new URL history file for the IP address;
        4) If a match was found, returns "Good"; otherwise
        returns "Bad".

    Modifications:

***********************************************************/

function InitAuth()
{
// Declare $uid and $pw as global so they can be used in the calling page.
    global $uid, $pw;

// Get the ticket sent to this IP address.
    $IPaddr = $_SERVER['REMOTE_ADDR'];
    $filename = "/data/www/uauth/specific/CL".$IPaddr.".dat";
    $F01 = fopen($filename, "r");
    $csticket = fread($F01, 80);
    fclose($F01);

// Loop through UserID-Password pairs trying to duplicate
//   the client-side ticket.
    $clticket = raw_param('hpccTicket');
    $filename2 = "/data/www/uauth/general/pwds";
    $F02 = fopen($filename2, "r");

  $result = "Bad";

    while (!feof($F02))
        {
        $uid = rtrim(fgets($F02, 80), "\n\r");
        if ($uid == "false" || feof($F02)) break;
```

```
            $pw   = rtrim(fgets($F02, 80), "\n\r");
            $teststring = Digest01a($csticket, $uid, $pw);
            if ($teststring == $clticket) {$result = "Good"; break;}
            $uidsave = $uid;
            }

    fclose($F02);

// If the login was successful, write the userID and password to
//   a file associated with the IP address and create a
//   history file for the IP address.
    if ($result == "Good")
        {
        $filename3 = "/data/www/uauth/specific/UP".$IPaddr.".dat";
        $F03 = fopen($filename3, "w");
        fputs($F03, $uid."\n");
        fputs($F03, $pw."\n");
        fclose($F03);

        $filename4 = "/data/www/uauth/specific/HS".$IPaddr.".dat";
        $F04 = fopen($filename4, "w");
        fputs($F04, date("l, F j, Y")." at ".date("h:i:s A (T)")." : <b>Initiated
        HPCC Logon Session</b>\n");
        fclose($F04);
        }

    return $result."|".$teststring."|".$clticket."|".$uidsave."|".$pw."|".$csticke
t."|".strlen($csticket);
}

function raw_param($name)
{
    return ini_get('magic_quotes_gpc')
        ? stripslashes( $_GET[$name])
        : $_GET[$name];
}

function Digest01a($tok, $ud, $pwd)
{
    $i = ord($ud[3]);
    $Totaldgn = 473291;
    $i = $i % 17; if ($i == 0) {$i=7;}

    for ($j=0; $j<=($i*strlen($tok)); $j++)
    {
        $k = $j % strlen($tok);
        $m = $j % strlen($ud);
        $n = $k % strlen($pwd);
        $temp = ord($ud[$m]) * ord($tok[$k]) - ord($pwd[$n]);
        if ($temp < 0) {$temp = 42;}
        $Totaldgn += $temp;
    }
    return $Totaldgn;
}
InitAuth();
?>
```

ContAuth.phpinc: Server-side code for continuing use of restricted pages

```
<?
/*********************************************************

      Function:           ContAuth
      System:             EGSC HPCC
      Date:               January 26, 2005
      Modification Dates: May 16, 2005
                          May 31, 2005
                          June 1, 2005
                          June 9, 2005
                          June 29, 2005

      Purpose and Use:

      This function carries out continuing authentication
      for all pages in the secure area of the Web site once
      the user has logged in. This function processes as
      follows:

      (1) Reads in from the CL... file the server-side
          ticket for the IP address involved;
      (2) Reads in from the UP... file the userID and
          password for the IP address involved;
      (3) Computes the digest of the client-side ticket;
      (4) Compares its computed digest with the digest
          received in the 'hpccTicket' input parameter
          and sets the result to "Good" if the two
          digests are equal and "Bad" if they are not.
      (5) If the comparison result is "Good", then
          the URL for the current page is added to the
          history list for this login session and IP
          address.
      (6) The result ("Good" or "Bad") is returned.

      Modifications:

      May 16, 2005: This function was modified to add
          URLs to the history list for a login session
          and IP address. URLs are added in reverse order
          so that when the history list is used, the
          most recently visited pages will be at the top
          of the list.

      May 31, 2005: The function was modified to log
          all page accesses by a user to a file specific
          to that user. The log file is permanent in the
          sense that it persists across sessions without
          loss of data, unlike the session-history files
          which are flushed each time a new session begins
          from a particular IP address (regardless of
          the HPCC user ID).
```

```
    June 1, 2005: A "chmod(...)" command was added
        to make the file with the user ID and
        password readable only by user Apache.

    June 9, 2005: Two lines were added to check for the
        existence of the ticket file and userid/password
        file for the client IP address, and to return
        the result "Bad" immediately if either file
        does not exist. This insures that attempted
        accesses to restricted pages will be redirected
        to a failure page and will not display
        potentially compromising information in error
        messages.

    June 29, 2005: The "raw_param" function was modified
        to look for the variable named by the parameter
        in both the $_GET and $_POST arrays.

***********************************************************/

function ContAuth()
{
// Make the variables $uid and $pw global so they can be used on
// the calling page.
  global $uid, $pw;

// Pre-set the return value (result) to "Bad" so it does not need to be
//   set multiple times.
  $result = "Bad";

// Get the client-side ticket for this IP address.
  $IPaddr = $_SERVER['REMOTE_ADDR'];
  $filename = "/data/www/uauth/specific/CL".$IPaddr.".dat";
    if (!file_exists($filename))
       {header('Location: http://egscbeowulf.er.usgs.gov/AccessFailure.php'); exit();}
  $F01 = fopen($filename, "r");
  $csticket = fread($F01, 80);
  fclose($F01);

// Get the digest which was passed as a parameter.
  $clticket = raw_param('hpccTicket');

// Get the userID and password for this IP address.

  $filename2 = "/data/www/uauth/specific/UP".$IPaddr.".dat";
    if (!file_exists($filename2))
       {header('Location: http://egscbeowulf.er.usgs.gov/AccessFailure.php'); exit();}
  $F02 = fopen($filename2, "r");
  if (!$F02) {return $result;} // Return "Bad" if fopen fails.
    $uid = rtrim(fgets($F02, 80), "\n\r");
    if ($uid == "false" || feof($F02)) return $result;
    $pw  = rtrim(fgets($F02, 80), "\n\r");
    if ($pw == "false" || feof($F02)) return $result;
  fclose($F02);
  chmod($filename2, 0600);
```

```
// Create a version of the $uid with no white space.
  $temparray2 = preg_split('[\s+]',$uid);
  $fuid = "";
  for ($jj=0; $jj<count($temparray2); $jj++) {$fuid.=$temparray2[$jj];}

// Compute the digest and compare it to the one passed as a parameter.
     $teststring = Digest01a($csticket, $uid, $pw);
     if ($teststring == $clticket) {$result = "Good";}

// If the result is "Good" add the URL of the current page to
//   the history file for this IP address and session.

if ($result == "Good")
    {
     $filename4 = "/data/www/uauth/specific/HT".$IPaddr.".dat";
     $filename5 = "/data/www/uauth/specific/HS".$IPaddr.".dat";
     $filename6 = "/data/www/uauth/specific/PM".$fuid.".dat";
     $urlstr = " : <a href=\"".$_SERVER['SCRIPT_NAME']."\" onclick=\"LinkOut(this)\"
>".$_SERVER['SCRIPT_NAME']."</a>\n";
     $F04 = fopen($filename4, "w");
     $F06 = fopen($filename6, "a");
     fputs($F04, date("l, F j, Y")." at ".date("h:i:s A (T)").$urlstr);
     fputs($F06, date("l, F j, Y")." at ".date("h:i:s A (T)").$_SERVER['SCRIPT_
NAME']." :: ".$_SERVER['QUERY_STRING']."\n");
     $F05 = fopen($filename5, "r");
     while (!feof($F05))
        {
         $tstr = fgets($F05, 400);
         if (strlen($tstr)>5) {fputs($F04, $tstr);}
        }
     fclose($F05);
     fclose($F04);
     fclose($F06);
     copy ($filename4, $filename5);
    }

// End and return.
  return $result;
}

function raw_param($name)
{
  $value1 = $_GET[$name];
  $value2 = $_POST[$name];
  $value = $value1;
//  $F200 = fopen("/data/www/uauth/general/debug", "a");
//  fputs($F200,$value1." -- ".$value2."\n");
  if (strlen($value2) > strlen($value1)) $value = $value2;
//  fputs($F200, "Value = ".$value."\n");
//  fclose($F200);

  return  ini_get('magic_quotes_gpc')
            ? stripslashes( $value )
            : $value;
}
```

```php
function Digest01a($tok, $ud, $pwd)
{
  $i = ord($ud[3]);
  $Totaldgn = 473291;

  $i = $i % 17; if ($i == 0) {$i=7;}

  for ($j=0; $j<=($i*strlen($tok)); $j++)
   {
      $k = $j % strlen($tok);
      $m = $j % strlen($ud);
      $n = $k % strlen($pwd);
      $temp = ord($ud[$m]) * ord($tok[$k]) - ord($pwd[$n]);
        if ($temp < 0) {$temp = 42;}
      $Totaldgn += $temp;
   }

  return $Totaldgn;
}
?>
```

JavaScript Include Files

<u>InitLogin.js: Client-side code for initial authentication</u>

```
/**********************************************************************

   Function Set:                  Initial Secure-Access Login
   File Name:                     InitLogin.js
   System:                        EGSC HPCC
   Date:                          January 26, 2005
   Modification Dates:            June 2, 2005

   Purpose and use:

    This is a set of two JavaScript functions used for logging
    into the EGSC HPCC for secure access to applications.

    The function "SaveAUvalues(thisform)" copies the values
    sent by the server (hpccTicket.value) and entered at the
    keyboard by the user (aus1.value, the user ID; and
    aus2.value, the pass phrase) so they will be available
    for later access; then it overwrites the user ID and
    pass phrase on the screen so neither will be visible
    or reusable following login.

    The function Digest01a(ticket, uid, pwd) is the JavaScript
    version of the digest function also used on the server
    side for securing communications. The use of digests
    allows authentication without either user ID or pass
    phrase being sent over the network in any direct form,
    plaint text or encrypted.

    Modifications:

    June 2, 2005: This function set was reformatted
       without substantive changes to either of the functions.

 **********************************************************************/

var UIDsave="";
var PWDsave="";
var TKNsave="";

function SaveAUvalues(thisform)
{
  TKNsave=thisform.hpccTicket.value;
  UIDsave=thisform.aus1.value;
  PWDsave=thisform.aus2.value;
  thisform.aus1.value="********";
  thisform.aus2.value="********";
  var TokDigestNumber = Digest01a(TKNsave, UIDsave, PWDsave);
  thisform.hpccTicket.value = TokDigestNumber.toString(10);
  return;
}

function Digest01a(ticket, uid, pwd)
```

```
{
  var i = uid.charCodeAt(3);
  var j, k, m, n, temp;
  var Totaldgn = 473291;

  i = i % 17; if (i == 0) {i=7;}

  for (j=0; j<=(i*ticket.length); j++)
   {
       k = j % ticket.length;
       m = j % uid.length;
       n = k % pwd.length;
       temp = uid.charCodeAt(m) * ticket.charCodeAt(k) - pwd.charCodeAt(n);
         if (temp < 0) {temp = 42;}
       Totaldgn += temp;
   }

  return Totaldgn;
}
```

<u>ContLogin.js: Client-side code for continuing use of restricted pages</u>

```
/*****************************************************************

    Function Set:                  Continuing Secure Access
    File Name:                     ContLogin.js
    System:                        EGSC HPCC
    Date:                          January 26, 2005
    Modification Dates:            Various
                                   June 2, 2005
                                   June 29, 2005
                                   August 28, 2005
                                   August 29, 2005


    Purpose and use:

    This is a collection of JavaScript functions used for continuing
    secure access to application pages on the EGSC HPCC following
    initial login to the EGSC HPCC.

    There are four sections in this JavaScript function set:

        (1) The Digest Function;
        (2) Functions for linking out from anchors;
        (3) Functions for linking out from other than anchors; and
        (4) Utility functions.

    These client-side functions must be used in conjunction with
    a set of server-side functions.

    Modifications:

    June 2, 2005: The function set was reformatted and rearranged
        without substantive changes to the functions themselves.

    June 29, 2005: A function was added for linking out from
        a form.

    August 28, 2005: A new global variable ("document.linkoutOKchk")
        and a new function ("OutLinkOKChk()") were added so that
        whenever a page is unloaded it will be possible to check
        whether one of the link functions contained in this
        code set was used. If one of these link functions was
        not used, then presumably the user inadvertently used the
        brower "back", "forward", or "refresh" buttons. To avoid
        having the authentication chain broken by these mistakes,
        the new variable and function allow the user to be alerted
        and sent back to the Applications Homepage if "back" was
        used. The button is not effective for the "refresh" button
        and the "forward" button won't be usable since the "back"
        button is not in use. The names of the first two linkout
        functions were changed from LinkOut() and LinkOut(,) to
        LinkOut1() and LinkOut2(,) respectively. Unlike C#, different
        parameter signatures in JavaScript do not lead to different
        functions; therefore, a distinction was needed.
```

```
    August 29, 2005: Having found that the change made yesterday
       works for the "back" button but not for the "refresh"
       button, I modified the text of the alert box asserted when
       the user clicks on a browser navigation button to provide
       a more accurate notice of what to expect.

*********************************************************************/

/*******************
  Global variable used to prevent inadvertent out-linking
  with the "back", "forward", or "refresh" browser buttons.
*******************/

  document.linkoutOKchk = "NotOK";
  window.onunload = OutLinkOKChk;

/*********************************************************************
   SECTION (1) : The Digest Function
*********************************************************************/

function Digest01a(ticket, uid, pwd)
{
  var i = uid.charCodeAt(3);
  var j, k, m, n, temp;
  var Totaldgn = 473291;

  i = i % 17; if (i == 0) {i=7;}

  for (j=0; j<=(i*ticket.length); j++)
    {
        k = j % ticket.length;
        m = j % uid.length;
        n = k % pwd.length;
        temp = uid.charCodeAt(m) * ticket.charCodeAt(k) - pwd.charCodeAt(n);
          if (temp < 0) {temp = 42;}
        Totaldgn += temp;
    }

  return Totaldgn;
}

/*********************************************************************
   SECTION (2) : Functions for linking out from anchors
*********************************************************************/

/*** Use this with 'onclick=' to link to a new page from an anchor
     when no parameters other than "hpccTicket" need to be passed. ***/
function LinkOut1(thisanchor)
{
  uid = opener.UIDsave;
  pwd = opener.PWDsave;
  tok = document.TicketForm.hpccTicket.value;
  var TokDigestNumber = Digest01a(tok, uid, pwd);
  var TokDigestString = TokDigestNumber.toString(10);
  var urlstr = thisanchor.getAttribute("href");
```

```
  var fullURL = urlstr + "?hpccTicket=" + TokDigestString;
  thisanchor.href = fullURL;
  document.linkoutOKchk="OK";
  return;
}


/*** Use this with 'onclick=' to link to a page from an anchor after
     a delay as specified by the argument "waitseconds"
     when no parameters other than "hpccTicket" need to be passed.
     Typically the delay is used when a server-side process should
     be given a few seconds before the new page is generated and sent. ***/
function LinkOut2(thisanchor,waitseconds)
{
  uid = opener.UIDsave;
  pwd = opener.PWDsave;
  tok = document.TicketForm.hpccTicket.value;
  var TokDigestNumber = Digest01a(tok, uid, pwd);
  var TokDigestString = TokDigestNumber.toString(10);
  var urlstr = thisanchor.getAttribute("href");
  var fullURL = urlstr + "?hpccTicket=" + TokDigestString;
  setTimeout('DummyFunc()',waitseconds*1000);
  thisanchor.href = fullURL;
  document.linkoutOKchk="OK";
  return;
}


/*** Use this with 'onclick=' to link and pass parameters to a new page
     from an anchor. ***/
function LinkOutP(thisanchor,AdditionalParameters)
{
  uid = opener.UIDsave;
  pwd = opener.PWDsave;
  tok = document.TicketForm.hpccTicket.value;
  var TokDigestNumber = Digest01a(tok, uid, pwd);
  var TokDigestString = TokDigestNumber.toString(10);
  var urlstr = thisanchor.getAttribute("href");
  var fullURL = urlstr + "?hpccTicket=" + TokDigestString;
  if (AdditionalParameters.length >1) {fullURL = fullURL + AdditionalParameters;}
  thisanchor.href = fullURL;
  document.linkoutOKchk="OK";
  return;
}

/***************************************************************************
   SECTION (3) : Functions for linking out from other than an anchor.
                 (These functions typically are referenced by
                 "onclick" in buttons or "onload" in the <body>
                 tag.)
   ***************************************************************************/


/*** Use this with 'onclick=' to refresh a page (link a page back to itself)
     from a button following a delay of the number of seconds specified
     by the argument "waitseconds". This function can be used (cautiously) in the
```

```
      <body> tag to cause a page to refresh itself periodically. ***/
function LinkOutRefresh(waitseconds)
{
  uid = opener.UIDsave;
  pwd = opener.PWDsave;
  tok = document.TicketForm.hpccTicket.value;
  var TokDigestNumber = Digest01a(tok, uid, pwd);
  var TokDigestString = TokDigestNumber.toString(10);
  var urlstr = location.pathname;
  var fullURL = urlstr + "?hpccTicket=" + TokDigestString;
  setTimeout('DummyFunc()',waitseconds*1000);
  self.location = fullURL;
  document.linkoutOKchk="OK";
  return;
}


/*** Use this with 'onclick=' to refresh a page (link a page back to itself) when the
     page has been called with parameters. This function is used from a button.
     It waits the number of seconds specified by the argument "waitseconds".
     This function can be used (cautiously) in the <body> tag to cause a page
     to refresh itself periodically. This function should not be used in a page
     that performs some initial process that should be performed only once. ***/
function LinkOutRefreshP(waitseconds)
{
  uid = opener.UIDsave;
  pwd = opener.PWDsave;
  tok = document.TicketForm.hpccTicket.value;
  var TokDigestNumber = Digest01a(tok, uid, pwd);
  var TokDigestString = TokDigestNumber.toString(10);
  var urlstr = location.pathname;
  var addpmt = location.search.substr(location.search.indexOf("&"));
  var fullURL = urlstr + "?hpccTicket=" + TokDigestString + addpmt;
  setTimeout('DummyFunc()',waitseconds*1000);
  self.location = fullURL;
  document.linkoutOKchk="OK";
  return;
}


/*** Use this function from a button (or other event trigger) to link and pass
     the current page's parameters to a new page specified by the argument
     "nurlstr" (new URL string). ***/
function LinkOutU(nurlstr)
{
  uid = opener.UIDsave;
  pwd = opener.PWDsave;
  tok = document.TicketForm.hpccTicket.value;
  var TokDigestNumber = Digest01a(tok, uid, pwd);
  var TokDigestString = TokDigestNumber.toString(10);
  var urlstr = nurlstr;
  var addpmt = location.search.substr(location.search.indexOf("&"));
  var fullURL = urlstr + "?hpccTicket=" + TokDigestString + addpmt
  self.location = fullURL;
  document.linkoutOKchk="OK";
  return;
```

```
}


/*** Use this with 'onclick=' to link to a new page from a form.  ***/
function LinkOutF(thisform)
{
  uid = opener.UIDsave;
  pwd = opener.PWDsave;
  tok = thisform.hpccTicket.value;
  var TokDigestNumber = Digest01a(tok, uid, pwd);
  var TokDigestString = TokDigestNumber.toString(10);
  thisform.hpccTicket.value = TokDigestString;
  document.linkoutOKchk="OK";
  return;
}



/*******************************************************************
   SECTION (4) : Utility functions
*******************************************************************/


/*** The HWinClose() function closes a browser window on the server side. ***/
function HWinClose()
{
  document.linkoutOKchk="OK";
  self.close();
}


/*** This function performs a client-side redirection to the "Logout" script
     which performs various housekeeping tasks on the server side when an
     EGSC HPCC user logs out of a secure application-usage session. ***/
function HWinJump()
{
  document.linkoutOKchk="OK";
  self.location = "http://egscbeowulf.er.usgs.gov/applications/Logout.php";
}


/*** This function is used with "setTimeout()" to provide a null operation
   so other functions may sleep without side effects. ***/
function DummyFunc()
{
   var dummyvar1 = 0;
}
```

```
/*** This function is an event handler for the Window (Page) "onunload"
     event. It prevents outlinking with the browser "back" button; it
     is not effective with the "refresh" buttons. Since "back" is
     not used, the use of the "forward" button is not an issue.       ***/
function OutLinkOKChk()
{
   if (document.linkoutOKchk != "OK")
       {
         uid = parent.UIDsave;
         pwd = parent.PWDsave;
         tok = document.TicketForm.hpccTicket.value;
         var TokDigestNumber = Digest01a(tok, uid, pwd);
         var TokDigestString = TokDigestNumber.toString(10);
         var urlstr = "/applications/AppHome.php";
         var fullURL = urlstr + "?hpccTicket=" + TokDigestString;
         var alertString = "Please remember NOT to use the browser navigation buttons
while logged in for secure access.\n\n";
         alertString = alertString + "If you just used your browser's \"back\" button
you will be returned to the Application\n";
         alertString = alertString + "Homepage. If you used the \"refresh\" button you
will be sent to an error page and you\n";
         alertString = alertString + "will have to log back in to resume your work.
(Sorry about that.)";
         alert(alertString);
         self.location=fullURL;
         return;
       }
   else
       {
         return;
       }
}
```

www.ingramcontent.com/pod-product-compliance
Lightning Source LLC
Chambersburg PA
CBHW081617170526
45166CB00009B/3006

* 9 7 8 1 5 0 0 2 9 7 5 4 1 *